The Best Study Series for
GED
Science
2nd Edition

Arthur R. Wagner, M.S.

Edited by
Jay Templin, Ed.D.

Research & Education Association
Visit our website at
www.rea.com

Research & Education Association
61 Ethel Road West
Piscataway, New Jersey 08854
E-mail: info@rea.com

**The Best Study Series for
GED SCIENCE**

Copyright © 2006 by Research & Education Association, Inc.
Prior edition copyright © 2000 under the title *The Best Pre-GED Study Series: Science*. All rights reserved. No part of this book may be reproduced in any form without permission of the publisher.

Printed in the United States of America

Library of Congress Control Number 2005929493

International Standard Book Number 0-7386-0126-8

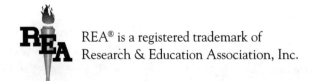

REA® is a registered trademark of
Research & Education Association, Inc.

TABLE OF CONTENTS

Introduction v
    About this Book vi
    About the GED Program vi
    About the GED Exam vii
    About the Science Test of the GED viii
    When Should I Start Studying? viii
    Before the Test viii
    Top Test Tips When Taking the GED viii
    After the Test ix
    About Research & Education Association x
    Acknowledgments x

Pre-Test 1
    Pre-Test 3
    Answer Key 7
    Pre-Test Self-Evaluation 8
    Pre-Test Answers and Explanations 10

Life Sciences 13
    Making Predictions 15
    Making Inferences 19
    Understanding a Diagram 23
    Getting Meaning From Context 27
    Summarizing Information 29
    Recognizing Cause and Effect 32
    Fact vs. Opinion 34
    Understanding a Photo 38
    Topics in Life Science 40
    Review 56

Chemistry 57
    Finding the Main Idea 59

Comparing and Contrasting 61
　　　Roots, Suffixes, and Prefixes 62
　　　Making Predictions 63
　　　Reading a Line Graph 65
　　　Topics in Chemistry 68
　　　Physical Science...71
　　　Review 75

Earth Science **77**
　　　Reading a Map 79
　　　Understanding Compound Words 92
　　　Finding Details...93
　　　Drawing Conclusions 96
　　　Reading a Circle Graph..98
　　　Reading a Bar Graph..102
　　　Topics in Earth Science..106
　　　Review 117

Physics **119**
　　　Recognizing Cause and Effect 121
　　　Making Inferences 124
　　　Reading a Diagram 126
　　　Drawing Conclusions 130
　　　Topics in Physics..134
　　　Review 140

Post-Test **143**
　　　Post-Test..145
　　　Answer Key 150
　　　Post-Test Self-Evaluation...151
　　　Post-Test Answers and Explanations 153

Appendix: Glossary of Terms **157**

Requirements for Issuance of Certificate/Diploma **175**

Science

Introduction

ABOUT THIS BOOK

Congratulations. You've taken an important step toward successfully earning your GED. This book is designed to help you take and pass the GED *Science Test*. Perhaps you've decided to focus your study time on the single subject that's most difficult for you, or you may have already taken the GED and did not pass completely, and you've chosen REA's single-subject guides for targeted review. Whichever applies to your individual situation, you've made a great choice.

You will begin your prep for the GED *Science Test* by learning what's expected of you on this exam. You will learn what topics are covered, how many questions to expect, and how much time you'll have to answer them. You will then move on to the subject review, where the subject matter will be discussed in much greater detail. The subject review in this book covers everything you'll need to know to pass the *Science Test*. Be sure to complete the drills in each section, since they will be a great help in keeping on top of your progress. If you find that you are struggling with the drill questions in a given section, it may be a good idea to read that section again.

The formula for success on the GED is very simple: the more you study, the more likely you are to pass the exam. Know the material covered here, and you will be calm and confident on exam day.

ABOUT THE GED PROGRAM

For more than 60 years, the GED Examination has been administered by the GED Testing Service of the American Council on Education (ACE). The GED exam offers anyone who did not complete high school the opportunity to earn a High School Equivalency Certificate. With that certificate come many new opportunities for a better career or higher education.

The GED may be a step on your journey to a college degree, since almost one out of every twenty first-year college students has a GED. Or a GED may be your ticket to a better job and into a career with a bright future and room to grow. Whatever your academic or professional goals are, success on the GED is a great place to begin.

The GED exam is available in all 50 states and Canada. There are over 3,400 testing centers in North America, and another 2,800 testing centers worldwide, so you should have no problem finding a GED testing center near you.

For more information on the GED program, to find an administration schedule, or to find a testing center near you, start by contacting your local high school or adult education center. Or, you can contact the American Council on Education, which administers the GED, at:

GED – General Educational Development
American Council on Education
One Dupont Circle NW, Suite 250
Washington, DC 20036-1163

To contact the GED administrators by phone, call: (202) 939-9300 or (800) 626-9433 (Toll Free)

You can also e-mail them at comments@ace.nehu.edu, or visit them on the Web at *www.gedtest.org*.

Alternate-Language GED Tests

If English is not your first language, you may be able to take the GED exam in Spanish or French. Contact the GED administrators for more information.

Accommodations for Test Takers with Disabilities

If you have special needs because of a physical or learning disability, accommodations are available for you. Some examples of qualifying disabilities are blindness, deafness, dyslexia, dyscalculia, and attention-deficit/hyperactivity disorder. A complete list of qualifying disabilities is available from the GED test administrators. The proper accommodations can make a great deal of difference for those entitled to them, so be sure that you are taking the GED exam that's right for you.

If you believe that you have a qualifying disability but you do not have complete documentation, contact the National Rehabilitation Information Center (NARIC) at (800) 346-2742.

ABOUT THE GED EXAM

The GED exam consists of five separate subject tests. The breakdown is outlined in the chart below.

The entire exam is 7 hours and 30 minutes long. That sounds like a lot to cover! But don't worry. Each topic is treated individually. If you pass all five topics in one sitting, you've earned your GED. If you don't, you only have to take the section or sections that you did not pass.

An Overview of the GED Tests

Test Area	Number of Questions	Time For Test	Test Format
Language Arts, Writing, Part I	50 questions	75 minutes	Organization (15%) Sentence Structure (30%) Usage (30%) Mechanics (25%)
Language Arts, Writing, Part II	1 essay	45 minutes	Written essay on assigned topic
Language Arts, Reading	40 questions	65 minutes	Literary Text (75%) Non-Literary Text (25%)
Mathematics	50 questions	90 minutes	Number operations and number sense (20%–30%) Measurement and geometry (20%–30%) Data analysis, statistics, and probability (20%–30%) Algebra, functions, and patterns (20%–30%)
Social Studies	50 questions	70 minutes	History (40%) Geography (15%) Civics/Government (25%) Economics (25%)
Science	50 questions	80 minutes	Physical Science (35%) Life Science (45%) Earth and Space Science (20%)

ABOUT THE SCIENCE TEST

This book will focus on the GED *Science Test*.

The *Science Test* presents you with 50 multiple-choice questions in the following content areas:

Physical Science (physics and chemistry) (35%)

Life Science (45%)

Earth and Space Science (20%)

Test questions require you to understand, interpret, or apply information that is provided on the test or that is learned through life experience. The information may appear in the form of a paragraph, chart, table, graph, map, or figure.

WHEN SHOULD I START STUDYING?

If you're wondering when to start studying, the short answer is *now*. You may have a few days, a few weeks, or a few months to prepare for the particular administration of the GED that you're going to take. But in any case, the more time you spend studying for the GED, the better.

BEFORE THE TEST

At some point, you've studied all you can and test day is only one good night's sleep away. Be sure to go to bed early on the night before test day, and get as much rest as you can. Eat a good breakfast. Dress in layers that can be added or removed so you'll be comfortable if the testing center is warmer or cooler than you like. Plan to arrive at the test center at least 20 minutes early so that traffic or other transportation issues don't prevent you from getting to the test center on time. If you're not sure where the test center is, be sure to make the trip at least once before test day. On the morning of test day, your only job is to let nothing—not hunger, not temperature, not traffic—distract you from your main goal: success on the GED. Use the test-day checklist at the back of this book to make sure you've covered all the bases.

What to Bring with You
- Your admission ticket, if you need one
- An official photo ID
- Some sharpened No. 2 pencils (with erasers) and a blue or black ink pen
- A watch, if you have one.

The following items will *not* be allowed in the testing area, so if you choose to bring them, know that you will have to store them during the test:
- Purses and tote bags
- Electronic devices, including MP3 players, video games, pagers, cell phones, CD players, etc.
- Food
- Books and notebooks
- Other non-essential items

Remember that by the time you reach test day, you will have put in your study time, taken your practice exams, and learned the test format. You will be calm and confident, knowing that you're ready for the GED.

TOP TEST TIPS WHEN TAKING THE GED

While you're taking the GED, here are some important test strategies:

- Read all the directions carefully so that you understand what's expected of you. If you have questions, ask the GED examiner.

- Answer every question. There's no wrong-answer penalty on the GED, so if you don't know, *guess*. If you leave a question blank, you're guaranteed to get zero points. If you have to guess, you have a 20% chance of getting the question right.

- Smart guesses are better than random guesses. If you have five possible answers, and you have no idea which is correct, that's a random guess. If you have five possible answers and you've eliminated three that are definitely wrong, that's a Smart Guess. You now have a 50% chance of getting the question right.

- Keep an eye on your time. Don't spend too much time on any one question. Choose your best answer, and move on. Come back to troublesome questions later, if there's time.

AFTER THE TEST

When you've completed your GED exam, you've reached the end of one journey and the beginning of another. You gave your best effort and it's a great feeling. Hopefully you've passed all five tests and you can look forward to a new world of opportunity. If you don't pass all five tests, you can focus your study time on only those areas that still need work. But remember, success begins with a goal, and whether you pass on your first try or not, your journey is well under way. Go home, relax, and take a well-deserved rest. You've earned it.

ABOUT RESEARCH & EDUCATION ASSOCIATION

Founded in 1959, Research & Education Association is dedicated to publishing the finest and most effective educational materials—including software, study guides, and test preps—for students in middle school, high school, college, graduate school, and beyond.

REA's Test Preparation series includes books and software for all academic levels in almost all disciplines. Research & Education Association publishes test preps for students who have not yet entered high school, as well as high school students preparing to enter college. Students from countries around the world seeking to attend college in the United States will find the assistance they need in REA's publications. For college students seeking advanced degrees, REA publishes test preps for many major graduate school admission examinations in a wide variety of disciplines, including engineering, law, and medicine. Students at every level, in every field, with every ambition can find what they are looking for among REA's publications.

REA's practice tests are always based upon the most recently administered exams, and include every type of question that you can expect on the actual exams.

REA's publications and educational materials are highly regarded and continually receive an unprecedented amount of praise from professionals, instructors, librarians, parents, and students. Our authors are as diverse as the fields represented in the books we publish. They are well-known in their respective disciplines and serve on the faculties of prestigious high schools, colleges, and universities throughout the United States and Canada.

We invite you to visit us at *www.rea.com* to find out how "REA is making the world smarter."

ACKNOWLEDGMENTS

Special recognition is extended to the following persons:

Larry Kling, Vice President, Editorial, for his overall direction.

Pam Weston, Vice President, Publishing, for setting the quality standards for production integrity and managing the publication to completion.

Stacey Farkas, Senior Editor, for project management.

John Paul Cording, Vice President, Technology, for his editorial contributions.

Christine Saul, Senior Graphic Artist, for designing our cover.

Jeremy Rech, Graphic Artist, for post-production file mapping.

Network Typesetting, Inc., for typesetting the manuscript.

Science

Pre-Test

SCIENCE

PRE-TEST

> **DIRECTIONS:** Carefully read and answer each of the following questions. Choose the <u>best answer choice</u> for each question.

1. _____ is the study of living systems and processes.

    (1) Astronomy

    (2) Meteorology

    (3) Biology

2. Which of the following is an example of a chemical reaction?

    (1) Burning a log

    (2) Letting air out of a tire

    (3) Hammering a nail

3. Which statement best describes "earth science?"

    (1) The study of the planet Earth in comparison to other planets in the solar system

    (2) The study of the history of science throughout the entire world

    (3) The study of the Earth's natural systems and structures

4. Which phrase is associated with the description of a cause-and-effect relationship?

    (1) Working on

    (2) Because of

    (3) Seeing through

5. When an object hits another object, scientists say a _____ has occurred between them.

    (1) cohesion

    (2) collision

    (3) attraction

6. Drawing conclusions is important to the understanding of earth science because _____.

    (1) natural events such as earthquakes and storms always end, no matter how powerful they are

    (2) one must conclude that Earth is made up of many interrelated systems

    (3) the ability to take information and draw a conclusion from it can bring about a greater understanding of the complicated systems and phenomena that comprise earth science

7. An example of an element is _____.
   (1) gasoline
   (2) wood
   (3) carbon

8. To _____ means to draw conclusions from statements you believe to be true.
   (1) infer
   (2) guess
   (3) know

9. The basic building blocks of matter are _____.
   (1) formulas
   (2) compounds
   (3) atoms

10. An easy way to present numerical information and figures is by using _____.
    (1) graphs
    (2) written interpretations of figures
    (3) formulas

11. A drawing that shows only the important parts of a system is a _____.
    (1) photograph
    (2) graph
    (3) diagram

12. Physics is the study of _____ and _____ moving about in space and time.
    (1) protons, electrons
    (2) rock, soil
    (3) matter, energy

13. The numerical naming of compounds using roots, suffixes, and prefixes can be used to determine _____.
    (1) the rate at which a compound naturally occurs
    (2) the number of atoms of a given element in a compound
    (3) the half-life of a radioactive element or compound

14. Maps can be used to present what kind of information?
    (1) The structure of a thunderstorm
    (2) The areas of the United States that are likely to experience thunderstorms on a given day
    (3) The different types of lightning that occur within a thunderstorm

15. Which answer best describes the relationship between science and compound words?
    (1) Compound words are not used in science
    (2) Compound words can combine two scientific concepts into a larger scientific concept
    (3) Compound words refer to chemical compounds

16. Which object is a compound machine?
    (1) Bicycle
    (2) Tire
    (3) Chain

17. Which word is associated with the following definition: "To reduce information into a few words or sentences that explain the main idea of a subject matter"?

(1) Enhance

(2) Summarize

(3) Expand

18. White paper slowly turns brown over time due to _____.

   (1) a bacteria contained in paper

   (2) invisible dirt molecules

   (3) a chemical reaction

19. Which subject matter is *not* considered to be part of "earth science"?

   (1) Weather

   (2) Earthquakes

   (3) Mathematics

20. An easily understandable method of picturing the relationship between two quantities is _____.

   (1) a line graph

   (2) an essay

   (3) a theory

21. Which term means "something follows logically from something else"?

   (1) Desire

   (2) Calculate

   (3) Infer

22. Hidden information can be recognized when studying a _____.

   (1) photograph

   (2) formula

   (3) graph

23. If a unit of distance on a diagram directly relates to a unit of distance in the real world, then we say the diagram is drawn _____.

   (1) to scale

   (2) by comparison

   (3) with a key

24. In science, when there is an effect there is an associated _____.

   (1) concept

   (2) cause

   (3) reaction

25. A(n) _____ is something that is provable and is widely accepted as being true.

   (1) opinion

   (2) fact

   (3) perception

26. A reaction between two elements and/or compounds is demonstrated by use of _____.

   (1) the Periodic Table of Elements

   (2) a calculator

   (3) a formula

27. Understanding details is an important aspect of the entire subject known as earth science because _____.

   (1) subtle details concerning the Earth's systems and composition can greatly affect the workings and interactions of the Earth's systems and composition of those systems

   (2) details are an important part of weather forecasting

(3) details specify the types of soil in the Earth's crust

28. If a scientific conclusion is made regarding the future, then it is called a _____.
    (1) prediction
    (2) summarization
    (3) review

29. A hypothesis explains and relates _____.
    (1) conclusions
    (2) facts
    (3) theories

30. Substances that are made of more than one element and have changed atomic bonds are called _____.
    (1) compounds
    (2) atoms
    (3) mixtures

31. The unit of measurement for latitude and longitude is _____.
    (1) miles
    (2) feet
    (3) degrees

32. The study of matter and energy moving about in space and time is _____.
    (1) biology
    (2) physics
    (3) chemistry

33. A formula will _____.
    (1) describe the organization and composition of a substance
    (2) tell the rate of a chemical reaction
    (3) describe the physical properties of each element involved

SCIENCE

ANSWER KEY

1. (3)
2. (1)
3. (3)
4. (2)
5. (2)
6. (3)
7. (3)
8. (1)
9. (3)
10. (1)
11. (3)
12. (3)
13. (2)
14. (2)
15. (2)
16. (1)
17. (2)
18. (3)
19. (3)
20. (1)
21. (3)
22. (1)
23. (1)
24. (2)
25. (2)
26. (3)
27. (1)
28. (1)
29. (2)
30. (1)
31. (3)
32. (2)
33. (1)

GED Science

PRE-TEST SELF-EVALUATION

Question Number	Subject Matter Tested	Section to Study (section, heading)
1.	inferences	II, Making Inferences
2.	predictions	III, Making Predictions
3.	details; conclusions	IV, Finding Details; IV, Drawing Conclusions
4.	cause-and-effect relationships	V, Recognizing Cause and Effect
5.	inferences	V, Making Inferences
6.	conclusions	IV, Drawing Conclusions
7.	main idea	III, Finding the Main Idea
8.	inferences	II, Making Inferences
9.	main idea	III, Finding the Main Idea
10.	graphs	IV, Reading a Circle Graph; IV, Reading a Bar Graph
11.	diagrams	II, Understanding a Diagram
12.	cause-and-effect relationships	V, Recognizing Cause and Effect
13.	numerical naming	III, Roots, Suffixes, and Prefixes
14.	maps	IV, Reading a Map
15.	compound words	IV, Understanding Compound Words
16.	cause-and-effect relationships	V, Recognizing Cause and Effect
17.	summarizing information	II, Summarizing Information
18.	predictions	III, Making Predictions
19.	details; conclusions	IV, Finding Details
20.	graphs	III, Reading a Line Graph
21.	inferences	V, Making Inferences
22.	photographs	II, Understanding a Photograph
23.	diagrams	V, Reading a Diagram
24.	cause-and-effect relationships	II, Recognizing Cause and Effect

II = Life Sciences   III = Chemistry   IV = Earth Science   V = Physics

Pre-Test

Question Number	Subject Matter Tested	Section to Study (section, heading)
25.	facts vs. opinions	II, Fact vs. Opinion
26.	formulas	III, Roots, Suffixes, and Prefixes
27.	details	IV, Finding Details
28.	conclusions	IV, Drawing Conclusions
29.	defining hypothesis	II, Making Predictions
30.	defining substances	III, Finding the Main Idea
31.	latitude and longitude	IV, Reading a Map
32.	defining physics	V, Recognizing Cause and Effect
33.	formulas	III, Roots, Suffixes, and Prefixes

II = Life Sciences   III = Chemistry   IV = Earth Science   V = Physics

PRE-TEST ANSWERS AND EXPLANATIONS

1. **(3)** Astronomy (1) is the study of the stars and the universe. Meteorology (2) is the study of weather phenomena. Biology (3) is the study of living systems and processes.

2. **(1)** Letting air out of a tire (2) and hammering a nail (3) do not involve changes in the chemical compounds of which they are comprised. Burning a log (1) does involve changes in its chemical compounds and, therefore, constitutes a chemical reaction.

3. **(3)** (1) and (2) do not accurately describe "earth science." They make no mention of the systems and structures that comprise the Earth's natural workings and composition.

4. **(2)** Answer (2) is a phrase that can be used to describe a cause-and-effect relationship. For example, "The flowers grew because of the seeds that were planted." The phrases in answers (1) and (3) cannot be used to describe a cause-and-effect relationship without adding another phrase, such as "because of."

5. **(2)** When an object hits another object, scientists say a collision (2) has occurred between them. Cohesion (1) is a term from chemistry which means the attachment of molecules to one another. Attraction (3) is a term that describes the tendency for certain objects to be drawn toward one another.

6. **(3)** Answers (1) and (2) are true. However, they do not comprehensively answer the question concerning the importance of drawing conclusions. Answer (3) broadly answers the question regarding drawing conclusions.

7. **(3)** Gasoline (1) and wood (2) are both compounds which are made up of multiple elements. Carbon (3) is an element which appears on the Periodic Table of Elements.

8. **(1)** To infer (1) means to draw conclusions from statements you believe to be true. To guess (2) has no definable basis from which conclusions or beliefs are formed. To know (3) does not require one to draw conclusions since you already know the information.

9. **(3)** Formulas (1) are used to describe chemical reactions. Compounds (2) are comprised of two or more elements which are comprised of atoms. Atoms (3) are the lowest level of structure, or building blocks, studied in basic chemistry.

10. **(1)** Graphs are used to present numerical information in an easy way. Written interpretations of figures (2) are not easily recognizable and take time to comprehend. Formulas (3) might describe how the figures are obtained, but do not actually present figures.

11. **(3)** A diagram (3) is a drawing that shows only the important parts of a system. A photograph (1) is an exact visual documentation of a scene. A graph (2) is used to display numerical information.

12. **(3)** Physics is the study of matter and energy (3) moving about in space and time. Protons and electrons (1) are terms that are associated with chemistry. Rock and soil (2) are terms that are associated with geology.

13. **(2)** The number of a given element in a compound (2) is represented through numerical naming by the use of roots, suffixes, and prefixes. The rate at which a compound naturally occurs (1) and the half-life of a radioactive element or compound (3) cannot be determined simply by understanding the numerical name of a compound.

14. **(2)** Answers (1) and (3) cannot be represented on a map. A written description and diagram would be more appropriate to describe the structure of a thunderstorm and the different types of lightning that occur within a thunderstorm.

15. **(2)** Answers (1) and (3) do not accurately describe the relationship between science and compound words.

16. **(1)** A bicycle (1) is a compound machine. It consists of simple machines which act in a physical manner upon each other. Since a tire (2) and a chain (3) do not consist of simple machines they are not compound machines.

17. **(2)** To reduce information into a few words or sentences which explain the main idea is the definition of summarize (2). To enhance (1) and to expand (3) refer to the broadening or increasing of a matter.

18. **(3)** Neither bacteria (1) nor invisible dirt molecules (2) account for the tendency of white paper to slowly turn brown. A chemical reaction (3) between paper and oxygen is responsible for the browning of white paper over time.

19. **(3)** Weather (1) and earthquakes (2) are both subjects related to earth science. While mathematics is used to describe some aspects of earth science, it is considered an entirely separate subject.

20. **(1)** A line graph (1) can clearly demonstrate the relationship between two quantities by presenting the relationship in a visual manner that can be quickly understood. An essay (2) or a theory (3) can take a great deal of time and effort to read and understand.

21. **(3)** Infer (3) is defined as something which follows logically from something else. To desire (1) simply means that one wants something. To calculate (2) means to mathematically determine an answer.

22. **(1)** Hidden information can be recognized when studying a photograph (1). A formula (2) and a graph (3) demonstrate precise ideas or data.

23. **(1)** If a unit of distance on a diagram directly relates to a unit of distance in the real world, then we say the diagram is drawn to scale (1). Something drawn by comparison (2) is drawn by an estimate, not a direct relationship. A key (3) is the part of a diagram that would contain the scale and other important information.

24. **(2)** In science, when there is an effect, there is an associated cause (2). A concept (1) could be used to explain a cause-and-effect relationship. A reaction (3) is not associated with all scientific effects.

25. **(2)** A fact (2) is something that is provable and widely accepted as being true. An opinion (1) is a personal belief. A perception (3) is a personal observation or comprehension.

26. **(3)** The Periodic Table of Elements (1) and a calculator (2) can be useful to the

understanding of a reaction between two elements and/or compounds. The actual reaction between two elements and/or compounds is demonstrated by a formula (3) which presents the pre- and post-composition of elements or compounds that react.

27. **(1)** Answers (2) and (3) both relate to specific branches of earth science. Answer (1) broadly defines the relationship between details and the entire subject matter known as earth science.

28. **(1)** A prediction is a scientific conclusion regarding the future (1). A summarization (2) is the reduction of information into a brief description. A review (3) is an examination of something after it has been performed or has occurred.

29. **(2)** By definition, a hypothesis is an educated guess based upon a set of facts or observations. Further experiments, or more facts, are used to determine whether the hypothesis is true or false.

30. **(1)** Substances that are made of more than one element are called compounds (1). Atoms are the basic building blocks of matter (2). A mixture occurs when a substance is produced by merely mixing one or more elements without changing the atomic bonds (3).

31. **(3)** The unit of measurement for latitude and longitude lines is degrees. There are 360 equal degrees in a full circle.

32. **(2)** Physics is the study of matter and energy moving about in space and time. Biology is the study of living systems and processes (1). Chemistry is the science of matter and its changes at the atomic level (3).

33. **(1)** A chemical formula is a detailed description of how the elements are organized in the substance.

Science

Life Sciences

SCIENCE

LIFE SCIENCES

MAKING PREDICTIONS

You may have heard it said of someone that they have "good genes." This usually means that they have a very healthy body and a strong resistance to infections. It was long believed that when a baby is born, these good qualities are passed on from the combination of a healthy mother and a healthy father. We shall learn that this is mainly true, but a baby's present and future health also has much to do with what its mother does between conception and birth.

Today, that part of science we call **genetics** has become a major factor in our world's future. From making "miracle" drugs to predicting whether some people are at greater risk of disease than others, the study of genetics is one area that will always be of great importance.

In order to learn how scientists make predictions, we will look at a particular way of asking questions and getting answers about the universe—from the smallest to the largest things. This method of investigation works well, not only in genetics and the life sciences, but in all areas of research.

Questions

1. What does the word "predict" mean in the paragraph above? Look up its meaning in the dictionary.

2. Can we predict if a baby will be healthy if both the mother and the father have "good genes?" Explain.

Answers

1. To state, tell about, or make known in advance, on the basis of special knowledge.

2. Yes, to an extent. However, there are other factors such as the mother's behavior during the baby's development.

Science is the process in which the real world is observed in an attempt to understand sequences and connections in nature. Ideas and discoveries are then translated by humans into what we call **technology**. Using a computer, turning on a lamp, and getting an X-ray are all examples of technological applications of science. To solve a problem, science and technology provide answers.

Observation and analysis are very important functions in the scientific world. The purpose of science is to make the mysterious understandable for all. The process that makes the real world understandable is that of critical thought. When carried out in a certain way, such critical thought is known as the **scientific method**. The scientific method is not just a body of knowledge, but is more a way of thinking—a logical way.

The five key steps in using the scientific method are:

1. Clearly state a question about Nature.

2. Gather as much known information as possible; that is, research the problem.

3. Make an educated guess at an answer that fits as much of the known information as possible. This educated guess is called a *hypothesis*.

4. Keep testing the hypothesis by a logical series of observations and experiments. The survival of any hypothesis depends on the strength of the data to support it.

5. Change the hypothesis, if necessary, based on the observed facts.

The *true* strength of the hypothesis is not so much the mass of evidence to support it, but rather the ability of the concept to withstand attack and criticism once the hypothesis is published. The interplay of ideas is fundamental to scientific thought and is welcomed by true scientists. The publication of papers and presentations at conferences allows mass exposure of new ideas to the scientific population. If the concept/idea can, over a period of time, withstand the pressures and scrutiny of the scientific community and still prove to be valid, then the hypothesis can be used to predict the outcomes of future experiments. Well-tested hypotheses go on to become major *theories*.

A good example of how a hypothesis grows to be able to make predictions was given about 140 years ago. Starting in 1850, an

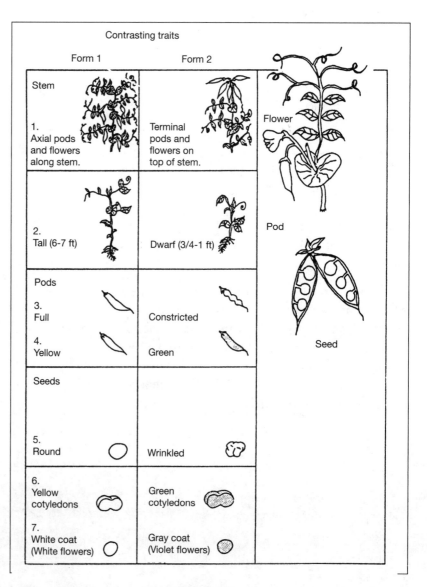

Figure 1

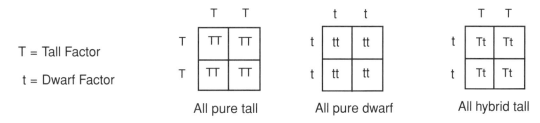

Figure 2

Austrian monk named Gregor Mendel began to study various offspring of common garden green pea plants. He studied seven characteristics, or traits, of garden peas as shown in Figure 1. After many trials and observations, Mendel found that when two tall plants are mated, they produce only tall offspring. The same held true for dwarf plants, which produced only dwarf offspring. However, when a tall plant was crossed with a dwarf plant, the result was always a tall plant. Mendel therefore noted that the pattern of inheritance in his peas featured a *dominant* and *recessive* relationship between alternate forms of each trait. For example, the tall trait is here shown to be dominant over the dwarfness trait.

The passing of traits from parents to their offspring is called **heredity**. From his many results over several years, Mendel was able to hypothesize that the factors of heredity occur in pairs. Mendel designated dominant factors with capital letters and recessive ones with corresponding lowercase letters. Thus, a pure tall plant would be designated TT, and a dwarf plant, tt (**homozygotes**). Those tall plants that were bred from TT and tt plants as parents were labeled Tt. They are known as **heterozygotes**.

By 1857, Mendel was able to show that his hypothesis of hereditary factors occurring in pairs always predicted results that were later observed. A diagram called a **Punnett square** makes the predictions exact. For example, a Punnett square for each of the three cases first studied for tallness vs. dwarfness appears in Figure 2.

We can go one step further and ask what Mendel's hypothesis of pairs predicts about the crossing of two tall heterozygotes. The predicted result is obtained from the Punnett square shown in Figure 3.

This predicts that there should be about three times as many talls as dwarfs from this union. In fact, Mendel's actual result was 787 tall plants and 277 short plants, roughly a 3:1 ratio.

In 1900, Walter Sutton expanded on Mendel's work by observing how certain rod-shaped bodies in cells called **chromosomes** relate to inherited traits. Sutton learned that the hereditary factors introduced by Mendel were units located in the chromosomes. He named these factors **genes** and the science of how traits are inherited **genetics**. Today, doctors can perform tests known as **genetic screening** to determine whether a person is going to be stricken in life by certain inherited diseases such as Huntington's disease or Down's Syndrome.

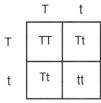

1/4 pure tall in offspring population

1/2 hybrid tall in offspring population

1/4 pure dwarf in offspring population

Figure 3

Questions

Are the following statements true or false? Circle your answer.

1. Gregor Mendel discovered chromosomes.
    T   F

2. Walter Sutton first used the term "genes."
    T   F

3. Controlled electricity is a form of technology.   T   F

Fill in the blanks with the best answers.

4. _____ is an inherited disease.

5. Chromosomes are found in _____ and are shaped like _____.

6. The diagram called a _____ helps to predict the ratios of offspring types.

7. _____ traits are ones that don't show up in heterozygotes.

Give answers to the following questions in the space provided.

8. Two hybrid (Rr) plants are crossed for the dominant trait (R) of round seeds. Out of 80 offspring, how many should have wrinkled seeds?

9. Someone wants to use the scientific method to cross the street safely. What are the five steps they take in doing this?

10. Name some inherited traits that people have.

_____ Match the following items with the given descriptions.

11. **Description**   **Item**

    : were shown by Mendel
    to occur in pairs   screening

    : are segments of   inheritance
    chromosomes   factors

    : is an educated guess
    about the answer   genes
    to a question

    : can detect inherited
    disease early   hypothesis

12. If a tall heterozygote pea plant is crossed with a pure dwarf pea plant, what should be the ratio of tall offspring to dwarf offspring?

13. If one of your direct relatives came down with an inherited disease, would you want to have genetic screening done to find out if you have it too? Briefly explain why.

Answers

1. False. Gregor Mendel discovered only that hereditary factors come in pairs.

2. True. Walter Sutton was the person who coined the word "genes."

3. True. Controlled electricity is a technological application of science.

4. Huntington's disease or Down's Syndrome

5. cells; rods

6. Punnett square

7. Recessive

8. Drawing a Punnett square,

	R	r
R	RR	Rr
r	Rr	rr

we predict that about 1/4 of the offspring, or about 20, will have wrinkled seeds.

9. To cross the street safely using the scientific method, a person:

   a. States the question clearly: "Can I cross the street safely?"

   b. Researches the problem and determines that, in order to cross safely, no dangerous vehicles must be where he or she is going to be while crossing.

   c. Make a hypothesis: "I can cross the street safely if I make sure to avoid all dangerous vehicles."

   d. Observes oncoming traffic from step to step, attempting to avoid all dangerous vehicles.

   e. Keeps the hypothesis if the street is crossed safely. Should an accident or near-accident occur anyway, modify the hypothesis to take into account why it happened for street crossings in the future.

10. Some inherited traits that people have are skin color, height, hair color, and eye color.

11. Inheritance factors were shown by Mendel to occur in pairs.

    Genes are segments of chromosomes.

    A hypothesis is an educated guess about the answer to a question.

    Screening can detect inherited disease early.

12. The Punnett square for a tall hybrid (Tt) plant crossed with a pure dwarf (tt) pea plant shows the following possible combinations:

	T	t
t	Tt	tt
t	Tt	tt

Thus, we see that for every one tall (Tt) offspring there should be one dwarf (tt) offspring. We say that their ratio is one-to-one, written as 1:1.

13. Individual answers may vary.

MAKING INFERENCES

Biology is the study of living systems and processes, and one of the main theories of biology is the **cell theory**. This theory states that (1) all living things are composed of cells, and (2) all cells arise from other cells.

There are many types of cells, such as plant cells, animal cells, and even one-celled animals! Cells also vary in size, although most of them fall into the range of 20 to 80 billionths of an inch long.

In this section, we will not only learn more about cells, but also about using what we learn to practice thinking. We will learn about *inferring* ideas from other things and how this is very useful in our everyday lives.

Questions

1. In the space below, write two things you believe to be true about cells:

   _____

   _____

   _____

2. Use a dictionary or an encyclopedia to find the numerical fraction for "one billionth." Write that fraction here:

   _____

Answers

1. All living things are composed of cells. All cells arise from other cells.

2. $\dfrac{1}{1{,}000{,}000{,}000}$

To **infer** means to draw a conclusion from statements you believe to be true. Look for connections and things that are implied by what you read or see or hear. That way, you can make *inferences* about them.

This is different from making predictions. A prediction deals with some time in the future. There is no idea of future time involved with an inference—although we could use it as part of a hypothesis to make predictions of the future.

For example, suppose we read the following information:

> The basic, self-sustaining unit of all life is called the cell. Most cells, whether plant or animal, have an outer container called the **cell membrane**. This membrane encloses a jelly-like substance called **cytoplasm** which contains organelles and the **nucleus**, the controller of the cell. The cell's chromosomes are found in the nucleus. These are Mendel's "factors of inheritance."
>
> The cell nucleus can be affected by radiation. Some forms of ultraviolet light can hit the nucleus and damage it. All cells reproduce to continue the process of life, but if the controlling nucleus is defective it may continue reproducing cells when it is not supposed to do so. This condition is known as **cancer**.

What can we infer from these facts? One thing could be that in order to avoid cancer from ultraviolet light we should avoid overexposure to the sun.

Another thing we could infer is that rocks are not alive because rocks are not made of cells.

Let's consider some more facts about living things:

Every living thing contains **proteins**. These are chains of molecules called **amino acids** which are chemically bonded together. Proteins are what make up our skin, our hair, our muscles, our cartilage, and many other components of our body. There is very little that is found in a living organism that is not composed, in part, of protein molecules. Protein molecules are specific to the organism of which they

are a part. Therefore, cow muscle protein cannot be found in the muscles of humans, and neither can chicken muscle protein be found in the muscles of hawks.

Now, suppose we were asked to complete the following multiple-choice statement:

Proteins are _____.

(1) chains of sugar molecules

(2) chains of fat molecules

(3) rings of carbon

(4) loops of cellulose molecules

(5) chains of amino acids

The correct answer is (5), but this is *not* an inference. Rather, it is a direct statement in the information given and not something implied by it. However, we *could* infer from the above that cats, dogs, horses, fish, and whales all contain protein because they are living things, too.

Read the following paragraphs and try to answer the questions by drawing inferences yourself:

A study of both plant and animal cells reveals the fact that in their most basic features, they are alike. However, they differ in several important ways. First of all, plant cells, but not animal cells, are surrounded by a rigid *cellulose* wall. This wall is actually a secretion of the plant cell itself. The wall surrounds the cell membrane and gives the cell its shape.

Another distinction between plant and animal cells is that many of the cells of green plants contain **chloroplasts**, which are not found in animals. These chloroplasts enable plants to make their own food from sunlight, water, and carbon dioxide.

Questions

1. Based on the information given, what can you infer about the shape of animal cells?

GED Science

2. What can you infer about an organism, like an animal, whose cells do not contain chloroplasts?

Answers

1. Animal cells, lacking a cell wall, do not have a constant shape.

2. Unlike green plants, animals need sources of food other than just sunlight, water, and carbon dioxide.

To sum up, making an inference means concluding that some concept or fact should logically follow from another. Remember that science is a process for understanding connections found in nature and, often, those connections must be inferred.

Questions

1. Are the following statements true or false? Circle your answer.

   a. Chromosomes are found in the cell nucleus.　　T　F

   b. Proteins contain amino acids.　　T　F

   c. An inference is the same as a hypothesis.　　T　F

2. Fill in the blanks.

   a. The unregulated growth of cells is called _____.

   b. _____ enable green plants to make their own food from _____, _____, and _____.

   c. Plant cells have a _____ that animal cells do not have.

3. Which of the following is the first barrier of defense that a foreign molecule must face upon entering an animal cell?

   (A) Nucleoplasm
   (B) Cytoplasm
   (C) Nuclear membrane
   (D) Chromosomes
   (E) Cell membrane

4. When animal cells divide, the cell surface constricts as if a belt were being tightened around it, pinching the old cell into two new ones. From what you know about plant cells, would you infer that they divide in the same way? Explain.

   _____
   _____
   _____

Answers

1. a. True; It was stated in the text, *The cell's chromosomes are found in the nucleus*. Therefore, it is inferred that chromosomes are found in the cell nucleus.

   b. True; It was stated in the text, *Every living thing contains proteins*. These are chains of molecules called amino acids that are chemically bonded together. Therefore, it is inferred that proteins contain amino acids.

   c. False; An inference is a conclusion from statements you believe to be true. There is no idea of future time involved with an inference, unlike a hypothesis. However, we could use an inference as part of a hypothesis.

2. a. cancer

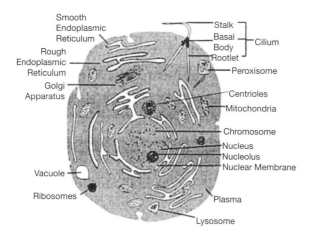

Figure 4: Typical Animal Cell

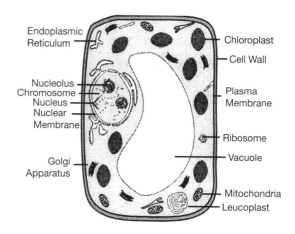

Figure 5: Typical Plant Cell

    b.   chloroplasts; sunlight; water; carbon dioxide

    c.   cellulose wall

3.   (E) You can infer that the cell membrane is the first barrier of defense a foreign molecule will face upon entering an animal cell since the cell membrane is the outer container of the cell.

4.   No, because plant cells have rigid cellulose walls which do not have the flexibility of animal cells.

UNDERSTANDING A DIAGRAM

There is an old proverb which says that "A picture is worth a thousand words." Today, a scientist might express the same thing by saying that "A picture has high information content." But no matter how it's said, the fact remains that one good picture of something does let us know a lot about it very quickly.

There are different kinds of pictures, such as drawings, paintings, and photographs. Some give us more information than others, depending on how many details are shown. It often happens that we cannot get or do not need a painting or photograph of something in order to study it. Even when these are available, we can sometimes add a diagram to them for further explanation.

A **diagram** is a drawing that shows only the important parts of a system. What is considered important depends on the level of study at the moment. Sometimes a rough sketch will

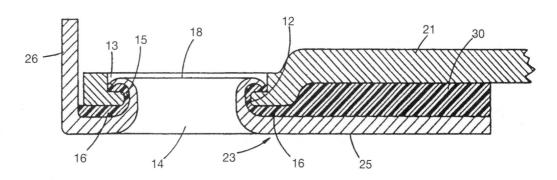

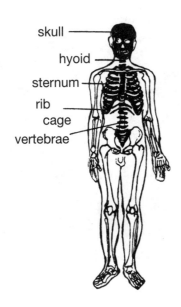
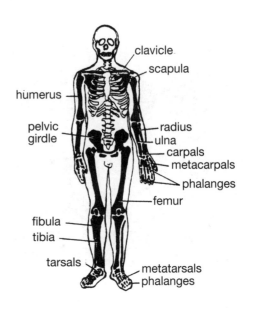

Figure 7a: The Axial Skeleton

Figure 7b: The Appendicular Skeleton

Questions

Write the titles of the three diagrams shown on page 23.

1. _____

2. _____

3. _____

Answers

1. Typical animal cell

2. Typical plant cell

3. Numbered parts diagram

do; other times, much more detail will be needed. A "good" diagram is one that gets the point across, no matter what the level of discussion.

Figures 4 and 5 on the previous page are diagrams of a plant cell and an animal cell.

The parts of any diagram are connected with identifying words called labels. In the cell diagrams, the parts and their labels are connected by lines.

In both Figures 4 and 5, circle the labels for the nucleus and the mitochondria.

We can also use the method of assigning numbers to the parts as labels and then explaining each part number elsewhere. This is the method used by the U.S. Patent Office and a typical example is shown in Figure 6 on the previous page.

The purpose of a **title** is to tell you the main content of the diagram. The title does not have to be at the top of the picture; in fact, it is usually underneath or off to the side.

The diagrams above show the bones in a person's body. Scientists divide the bones into two groups: those along the main axis of the spine (the axial skeleton) and all other bones (the appendicular skeleton). The word appendicular comes from the root word "append" meaning "in addition to."

The base of the axial skeleton forms a triangle called the sacrum, consisting of five fused vertebrae. From its end a number of small bones extend to form the coccyx.

The hyoid bone supports the tongue and its muscles. The last two ribs, also known as "floating ribs," have no attachment to the central breastbone, the sternum.

Questions

1. In the diagram of the appendicular skeleton shown in Figure 7b, where is the clavicle?

2. Which of the following is not a part of the axial skeleton?

    (1) Skull

    (2) Humerus

    (3) Ribs

    (4) Sternum

Answers

1. The clavicle is located above the scapula.

2. (2) The skull, ribs, and sternum are part of the axial skeleton. The humerus is part of the appendicular skeleton.

A place where two bones meet is called a joint. Some joints, such as those between the bones of the skull, cannot move and are very strong. The elbow and knee are hinge joints which permit movement in only one direction, while the hip is a ball-and-socket joint which permits movement in several directions.

The pivot joints at the wrists and ankles allow freedom of movement in an amount between that of the hinge and the ball-and-socket.

There are three types of muscles in humans: skeletal, cardiac, and smooth. All muscle tissue exerts force when it contracts; therefore, muscles are responsible for all movement of the body, voluntary and involuntary. The skeletal muscles are paired to accomplish full movement. Each contracting muscle will be paired with an antagonistic (i.e., opposing) muscle, and tendons attach paired muscle groups to bones to complete the movement action.

Cardiac muscle is found only in the heart. The heart is the strongest muscle of the body. It is responsible for keeping the blood flowing through the veins and arteries. A cardiac muscle is an involuntary muscle (involuntary muscle cannot be consciously controlled). A diagram of the human heart is shown in Figure 8.

Arteries are blood vessels which carry oxygenated blood away from the heart and

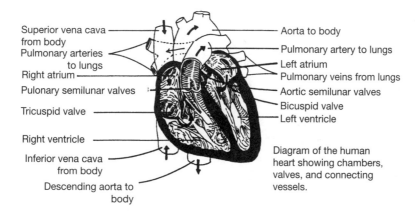

Figure 8: The Human Heart

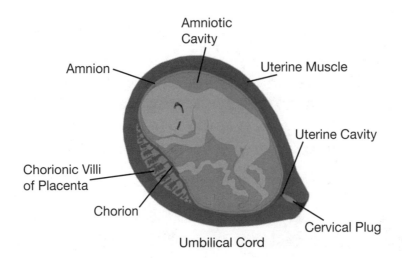

Figure 9: A Fetus

veins carry deoxygenated blood back to the heart. The pulmonary veins are unlike most veins in that they carry oxygenated blood.

In the diagram of the heart, circle and identify the pulmonary veins. When the coronary arteries become blocked with a substance called cholesterol plaque, not enough blood reaches the heart and part of it may die. This is what happens in a heart attack.

Finally, let us look at a diagram dealing with human reproduction shown in Figure 9.

In order to get to the stage of development shown in Figure 9, a sperm from the father had to unite with an egg from the mother to form a **zygote**. This fertilized egg then divided over and over to create a hollow round ball which, within ten days, attached itself to the wall of the mother's **uterine cavity**, or womb.

Between three and eight weeks, the growing baby is referred to as an **embryo**. The **placenta** is an organ that is made of both tissue from the mother and tissue from the baby. It regulates the passage of nutrients to and wastes from the embryo but never allows the blood of the mother to mix with the blood of the baby. Birth defects are most likely to occur while the embryo is developing the main organs of the body. All of this activity takes place within a separate enclosure in the womb called the **amniotic cavity**. It is formed by two membranes called the **amnion** (inner membrane) and the **chorion** (outer membrane).

From three to nine months of development, the baby is called a **fetus**. A tube about 70 cm long and 1 cm in diameter forms from the navel of the fetus to the placenta. This flexible tube is called the **umbilical cord**.

Questions

1. Write the main idea of the diagram (Figure 9).

   _____

   _____

   _____

   Fill in the blanks.

2. A diagram is a _____ that shows only the _____ parts of a system.

3. The words that identify the parts of a diagram are called _____.

4. The main idea of a diagram is usually stated in its _____.

5. The long, flexible tube through which the embryo receives nourishment from the uterine wall is the _____.

Refer to the diagram of the fish shown below.

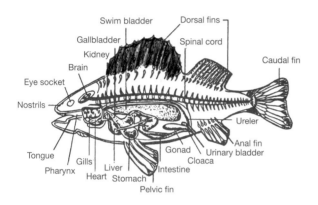

Diagram of a Fish

6. The fish's heart is closest to the (a) front (b) middle (c) rear of its body. _____

7. How many fins does the fish have? _____

8. The spinal cord starts at the _____ and ends at the _____.

9. The mouth cavity of the fish is called the _____.

Answers

1. The main idea of the diagram is the advanced fetus membrane and its relationship to the uterus.

2. drawing, important

3. labels

4. title

5. umbilical cord

6. (a)

7. Five, the fish has two dorsal fins, a caudal fin, an anal fin, and a pelvic fin.

8. brain, caudal fin

9. pharynx

GETTING MEANING FROM CONTEXT

When reading a passage, it often happens that you will come across a word or phrase whose meaning is new to you. However, you may be able to determine its meaning by how the unknown word is used. The information given before and after such a word or phrase is called its **context**. In this section, we will practice the skill of getting meaning from context while studying about tiny living things called microorganisms. Sometimes, we call these microorganisms germs.

Let's read the following passage and see what it has to offer:

> Germs are small, living creatures that get into plant or animal tissue for nourishment. By this process, some germs cause infectious diseases, and some infectious diseases are contagious. Germs are always present on the human body and can enter it in several ways such as breathing, eating, or through cuts in the skin.

Perhaps the word infectious is new to you. Can you interpret its meaning by other information around it—in context? Yes, because the sentence just before it says that it is in the nature of germs to get into plant and animal tissue, you could guess that infect means "to get into." An infectious disease, therefore, comes from a germ causing problems after getting into the system.

Do you know the meaning of the word contagious? If not, look it up in a dictionary. Do this with any words you are not sure of, or those whose meaning cannot be determined from their context.

There are many categories of microscopic organisms. These include protozoans, yeasts, algae, bacteria, and viruses. Many one-celled organisms are difficult to classify as either animals or plants. Some have characteristics of both animals and plants. One example of this type is called **Euglena**. In contrast, other organisms like bacteria are very different from both plants and animals. Like plants, bacteria have a cell wall, but this wall is not composed of cellulose. Some bacteria get energy from photosynthesis, as do plants, while others live on organic material the way animals do. A third kind even lives on inorganic compounds which they combine with oxygen.

From the context of the above information, we may conclude that Euglena is a microorganism and that it contains things like chloroplasts found in plants to make food by means of photosynthesis.

Technically speaking, a *virus* is not a true cell and is much, much smaller than a bacterium. Viruses infect a cell because they need the material in the cell to replicate and spread throughout the environment. Once attached to the *host*, a virus penetrates it and injects its own DNA. The viral DNA makes its way through the cytoplasm and into the nucleus where it uses the chromosomes to replicate. The new *virions* may escape from the host by causing a *lysis* of the host, but in other cases the host wall need not burst open to allow the release of the virions.

From context, we can say that: to replicate means to reproduce others like the originals; that a host is a cell that is invaded by a virus; that virions are new virus particles replicated in a host; and that lysis is the bursting of a host cell's wall by some types of virions in order for them to escape.

Questions

1. Define the meaning of the word "context" (use a dictionary if necessary):

   _____

   _____

2. Write a word in this section whose meaning was derived from context:

   _____

3. Read the following passage:

   Materials move into and out of a plant cell in order for it to stay alive and function. All transport occurs through and across a cell membrane. A certain amount of water pressure is needed inside the cell to adjust its requirements to its surroundings. It is by maintaining this turgor pressure that all movement through membranes of the cell is controlled.

   (1) In this passage, what is the meaning from context of the word transport?

   _____

   _____

(2) What is the meaning from context of the word turgor?

_____

_____

4. If a person is found to have AIDS (Acquired Immune Deficiency Syndrome), he or she may not be able to form *antibodies* against diseases like pneumonia. What does the word *antibodies* probably mean from context here?

_____

_____

Check your answer with a dictionary or an encyclopedia.

5. Protein degradation results in a collection of amino acids. These smallest protein units are then reconstructed into peptide chains by the process known as protein synthesis. From context, degradation means _____ and synthesis means_____.

Answers

1. Context means the setting or background in which a word is used.

2. Infectious; if you read the first passage on page 27, you can determine the meaning from context of the word "infectious." Read the paragraph that follows the passage to understand how to determine the meaning from context.

3. (1) The meaning from context of the word transport is the movement of material into and out of a cell.

(2) The meaning from context of the word turgor is a water regulator maintaining the cell pressure.

4. From the context provided, **antibodies** are disease-fighting cells.

5. destruction, creation

SUMMARIZING INFORMATION

It is often important, in any line of work, to be able to *summarize* information. This means being able to put it into a few words or sentences that get across the main ideas. You are trying to leave out as much detail as possible while still keeping the most important points.

In practicing this skill, ask the questions "What general things are being talked about?" and "What general happenings are being talked about?" This is how you make your summary short and yet describe the information itself. There may be several possible ways to state a summary but, however it is done, it must be both short and correct. Let's try practicing this skill on a reading from plant science:

A plant contains organs arranged in two systems: the root system and the shoot system. The shoot system is that portion of the plant that is above the ground, and the root system of the plant is that portion that is usually under the ground. The root system functions as an anchor and absorbs water and minerals from the ground. The stem supports the leaves and the flowers. The points where the leaves grow out of the stem are called nodes, and the space between two nodes is called an internode. Unlike the stem, roots have no internodes or nodes. The leaves function in photosynthesis while the flowers function as reproductive organs that ultimately produce seeds.

Ask yourself, "What general things are being discussed here?" The answers are (a) the shoot system above ground and (b) the root system below or near the ground, in any plant. The first supports leaves for photosynthesis and flowers for reproduction while the second serves as an anchor and a food carrier. Therefore, we can make the following summary from the information above: "Any plant is composed of two main systems—roots near or under the ground that serve as an anchor and nutrient carrier, and the shoots above ground having a stem with leaves for photosynthesis and flowers for reproduction. Let's try to summarize another reading:

A flower is composed of several different modified leaves (see Figure 10). The sepals are modified leaves that encircle the flower when it is in "bud." Sepals are green and protect the bud while it is developing into a flower. Just inside the sepals are the petals. The petals are usually large, showy, and brightly colored. They serve as a means of advertisement for animal pollinators that are attracted to bright patterns. Inside the petals are the stamens. The stamen is a two-part structure made up of the anther and the filament. The filament supports the anther. The anther is a structure that houses the developing pollen grains. In the very center of the flower is the pistil. The pistil is a series of modified leaves (the carpels) that contain the ovules. A pistil is composed of three basic parts: the stigma, the style, and the ovary/ovaries. The stigma is a sticky knob that receives the pollen. The pollen germinates and grows downward into a long and slender style and into the ovary and ultimately fertilizes the egg that is contained within the ovary.

We can draw a "connection tree" between general things and happenings as shown on the following page.

Our summary can be: "What we call a 'flower' is really a cluster of special leaves. Some (sepals) protect the forming bud; others (petals) attract animal pollinators, like birds and insects. Modified leaves known as the stamen form the male organ while others called the pistil make up the female reproductive organ."

Gypsy moths have been a great source of annoyance and tree damage in the northeastern United States for over one hundred years. When a Frenchman

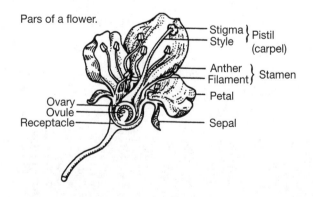

Figure 10: Parts of a Flower

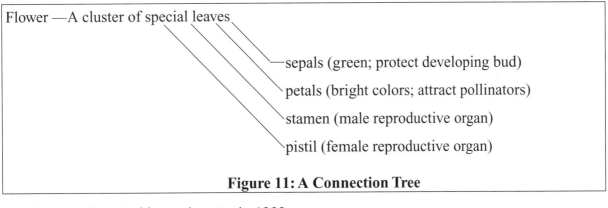

Figure 11: A Connection Tree

brought them to Massachusetts in 1860, a few of them escaped and today they are a major problem as far west as Michigan and as far south as Virginia.

Like many insects, the gypsy moth goes through several stages of development in what is called its **life cycle**. The life cycle of the gypsy moth occurs in four stages: the *egg*, the *caterpillar*, the *pupa*, and the *adult*. Each summer, a female moth lays several hundred eggs in a cluster. These eggs take about ten months to hatch into tiny caterpillars about one-quarter inch long. After one more month, the caterpillar has eaten through the leaves of every oak, birch, aspen, and gum tree it comes across and has grown to a length of about two inches. At this time, the caterpillar encloses itself in a casing for two weeks while it is changing form again. This is called the pupa stage. When it finally emerges from this shell, the animal is an adult gypsy moth with wings. Adult moths do not live long; they do not eat any food but only exist to mate and reproduce. The male dies after mating and the female dies after laying her eggs.

Pick out the key concepts from the previous two paragraphs concerning the gypsy moth. On a separate piece of paper, draw your own "connection tree."

Based on the connection tree, write a short summary of the paragraphs:

_____

_____

_____

Questions

Fill in the blanks.

1. To summarize means to _____.

2. The drawing that helps in summarizing ideas is called a _____.

3. The _____ of a plant keep it anchored.

4. The flower is a set of modified _____.

5. An organism's main stages of development are called its _____.

Match the following lists

A.	B.
6. Flower's (male) reproductive organ	(A) caterpillar
7. Eats tree leaves	(B) pistil
8. Encased stage in a life cycle	(C) stamen
9. Flower's (female) reproductive organ	(D) sepals

10. Protects buds   (E) pupa
11. Attracts animal pollinators   (F) petals

Answers

1. state in short words or sentences the general idea

2. connection tree

3. roots

4. leaves

5. life cycle

6. (C)

7. (A)

8. (E)

9. (B)

10. (D)

11. (F)

RECOGNIZING CAUSE AND EFFECT

Any time it is possible to make something happen by doing something else, scientists say that there is a cause-and-effect relationship between those two things. For example, lifting a weight off the ground and then letting it go is the cause of its falling back down—the effect. Such cause-and-effect relationships occur throughout nature and the area of living things is no exception. Plants and animals are closely connected to each other and disturbing the surroundings of these organisms affects the environment. There may be important connections to human beings down the line and we would do well to know about them.

When living things and nonliving things interact within a certain area on the Earth, scientists refer to that area as an **ecosystem**. Large museums and public aquariums usually display various kinds of ecosystems—the beach, the salt marsh, the tropical rainforests, the polar ice regions, and so forth.

In all ecosystems, plant and animal wastes are used to sustain other plants and animals. In the balance of nature, animals breathe oxygen, which is the waste gas from plants, and release carbon dioxide. Plants absorb carbon dioxide, an exhalation waste from animals, and use it through a process known as photosynthesis to sustain their existence. Oxygen is released by plants as a waste product during photosynthesis. Thus, plants and animals form an interlocking system, with plants absorbing carbon dioxide and releasing oxygen, and animals breathing oxygen and releasing carbon dioxide. Plants also extract nutrients from fecal matter and urine deposited by animals.

Predatory animals feed off of other animals (prey) to survive. Predator animals are

larger, faster, and generally smarter than prey animals. Prey animals maintain the existence of their species by reproducing at a higher rate.

Water plays an important part in the stability of the ecosystem. The sun causes water to evaporate and the water eventually falls to the Earth as rain, which provides an important resource for plants and animals. Of course, if an area receives little or no rain, it develops a desert ecosystem. Plants and animals which can be found in a desert ecosystem include: sidewinder rattlesnakes, gila monsters, kangaroo rats, cacti, tarantulas, black widow spiders, and scorpions. Cacti have tough skin that holds in moisture and needles that keep away animal predators. Cacti also have very deep and efficient root systems that capture and utilize the little rain which falls in a desert ecosystem. Animals give back water to the desert ecosystem in the forms of sweat and urine.

Flowering plants and pollinating insects have a mutualistic relationship in which they rely upon one another. Bees, wasps, moths, beetles, ants, etc. derive food from the pollen and nectar in plants. In return, these insects play an important role in the reproduction process of plants. Insects help to pollinate or fertilize plants through their constant interaction with flowers (the plants' reproductive organs). Without insects, plants would have to pollinate an opposite gender flower on the very same plant or rely on the luck of the wind for fertilization.

In arctic ecosystems, fungus-algae symbiotic life forms exist called lichen. The combination of fungus and algae is the only plant life hardy enough to live in arctic regions. Animals, such as reindeer, eat the lichen to survive in arctic regions.

Where specifically does the oxygen come from that we breathe? A large percentage comes from luxuriant tropical rainforests, which have an abundance of trees and plants. The majority comes from tiny ocean plants called phytoplankton that photosynthesize oxygen from the carbon dioxide that they absorb.

Another part of the global ecosystem is the stratospheric ozone layer or ozonosphere. Ozone is the combination of three oxygen molecules. The ozone layer protects plants and animals from dangerous and harmful ultraviolet rays from the sun. Ultraviolet rays can cause cancer in animals, including humans, and can damage the growth of plants. The ozone layer is being depleted by man-made chemicals called chlorofluorocarbons.

Geologic evidence shows that great floods, meteor impacts, and ice age glaciations and meltings have greatly affected the earth's ecosystem and have at times caused mass extinctions of plant and animal life.

Global warming has increasingly been viewed as a threat to the Earth's ecosystem. Global warming is theoretically being caused by a buildup of greenhouse gases, primarily carbon dioxide. The greenhouse gases trap the sun's rays in the Earth's atmosphere and cause the atmosphere (air) to warm. A certain amount of greenhouse effect is critical to the survival of the earth's ecosystem. Without any greenhouse effect, the Earth would be a much colder planet, unable to support the kinds of life that currently populate the Earth.

Insects are the most frequent and diverse species of multicellular organisms on Earth. The most common insect is the beetle. There are nearly one million species of beetles on earth. Half of these beetle species live in tropical rainforests where they are in great danger of extinction due to mankind's clear cutting of the rainforests and draining of wetlands. The effect of man's actions is the alteration of the ecosystem and the extinction of species such as beetles.

Questions

Match the following cause-and-effect relationships:

A. Animals/Oxygen   (1) Mass Extinctions

B. Predators   (2) Plants/Carbon Dioxide

C. Evaporation   (3) Oxygen

D. Flying Insects   (4) Global Warming

E. Tropical Rainforests/Phytoplankton   (5) Prey

F. Chlorofluorocarbons   (6) Plant Reproduction

G. Meteor Impacts   (7) Cutting of the Rainforest/Draining Wetlands

H. Increasing Carbon Dioxide Gas   (8) Ozone Depletion

I. Beetle Extinction   (9) Rainfall

Answers

A. (2) Plants absorb carbon dioxide and release oxygen. Animals breathe oxygen and release carbon dioxide.

B. (5) Predators hunt prey in order to survive.

C. (9) The sun causes water to evaporate, and the water eventually falls to the earth as rain.

D. (6) Flying insects (for example, bees) derive food from the pollen and nectar in plants. In return, insects play an important role in the reproduction of plants by pollinating and fertilizing plants through their constant interactions with flowers (plants' reproductive organs).

E. (3) A large percentage of oxygen comes from tropical rainforests. The majority of oxygen comes from phytoplankton, tiny aquatic plants, that photosynthesize oxygen from the carbon dioxide they absorb.

F. (8) Chlorofluorocarbons (man-made chemicals) are causing the depletion of the ozone layer.

G. (1) Meteor impacts have greatly affected the earth's ecosystem and have at times caused mass extinction of plant and animal life.

H. (4) Greenhouse gases, primarily carbon dioxide gas, cause global warming. Carbon dioxide, as well as other greenhouse gases, traps the sun's rays in the earth's atmosphere and causes the atmosphere (air) to warm.

I. (7) Beetle extinction is caused by man's cutting of the rainforest and draining of wetlands.

FACT VS. OPINION

Distinguishing what is considered to be a fact from an opinion is an important scientific skill. A fact is something that is provable and is widely accepted as being true. An opinion is a personal belief that is based upon personal observations and assumptions, which cannot be proven through experimental means. Scientific opinions eventually are accepted as fact or are disproved by direct experiment or observation. The theory of evolution provides us with an interesting discussion topic to learn how to distinguish scientific opinion from scientific fact. Carefully read the following passages regarding the development of the theory of evolution

Life Sciences

and the theory about the dinosaurs' extinction. The questions which follow each passage will help to develop skills which can be used to distinguish fact from opinion.

Since the time of the ancient Greeks, scholars have speculated on the development and origin of life. Many different opinions have been held through the ages. Expedocles, and later Lucretius, were among the first to suggest that more complex life forms evolved from simpler life forms. Other scholars, like Buffon and Paley, believed that life forms were pretty much the same through eternity, due to a built-in regularity of design. Lamarck believed that species would change when they needed to evolve new organs, and that these acquired characteristics could be inherited.

Charles Darwin said his original beliefs were similar to those of Buffon and Paley. Darwin changed his beliefs after he sailed on the *H.M.S. Beagle* to South America and the Galapagos Islands. He learned that great amounts of evolution occurred in species that he came across while on his journey. The Galapagos Finches were all adapted to different niches of island behavior and eating habits and yet they had evolved from mainland Ecuadorian Finches that had flown there. After reflection on the writings of Malthus on overpopulation, Darwin realized that the mechanism of evolution had to be natural selection. That is, those variations of the species that fit a particular environment would survive, and those that were unfit would die and keep the population down.

Pigeons and dog breeders supplied further evidence that natural selection could drive evolution since many subspecies or varieties of pigeons or dogs could be created from the variations of only a few types. This artificial selection and selective breeding worked very quickly and effectively and thus showed what nature was capable of through the process of natural selection.

Questions

Fill in the blanks.

1. A fact is something that is _____ and is widely accepted as being _____.

2. An opinion is a personal _____ that is based upon personal observations and assumptions, which cannot be proven through _____ means.

Answer the following questions regarding the preceding passage.

3. Were Expedocles' and Lucretius' statements about evolution facts or opinions? Explain.

   _____
   _____
   _____

4. What event caused Darwin to change his ideas about the origins and development of life? Explain.

   _____
   _____
   _____

5. Were Darwin's ideas about the origins and development of life facts or opinions when he first proposed them? Explain.

   _____

GED Science

6. Did the pigeon and dog breeding experimentation validate Darwin's ideas as facts, at least on a basic scientific level? Explain.

Answers

1. probable, true

2. belief, experimental

3. They were opinions. They did not cite specific proven evidence to support their statements.

4. When he traveled, Darwin realized that finches in regional climates adapted to their surroundings.

5. They were opinions when he first proposed them. His ideas were not originally widely accepted and there was little experimental evidence.

6. Yes, they proved that many varieties of a species can be linked to a few original types.

Read the passage and answer the questions that follow.

By the early 1800s scientists had obtained enormous, nearly complete, skeletons of extinct lizards. These creatures were labeled dinosaurs, which is derived from the Latin words for "terrible lizard." A great mystery arose from the fact that dinosaurs were extinct. Many opinions arose concerning the reasons why dinosaurs became extinct. Everything from massive volcanic eruptions, radical climate change, a meteorite impact on the Earth, to the eating of the dinosaurs' eggs by mammals was proposed as the reason for the dinosaurs' extinction. It was only with some luck and exploration that observations were made that supported some of the opinions, and the opinions were accepted by the scientific community as fact.

Geological scientists tell us that 65 million years ago all of the dinosaurs quickly died and the species became extinct. Until recently, there was no hard evidence about the cause of the dinosaurs' extinction. Unprovable opinions regarding the dinosaurs' extinction expanded and splintered and the mystery deepened. A major scientific breakthrough was required to break this legacy of never-ending opinions. This is exactly what occurred. Geologists observed that a 65 million-year- old, thin layer of iridium enriched rock was observable in rocks worldwide. The amount of iridium observed was too great to have originated from a source on the Earth, since almost all of the iridium on earth is trapped in the molten core. However, scientists observed that the amount of iridium found in the 65 million-year-old layer of rock matched the amount that can be expected from meteorites from outer space. As a mere opinion, the possibility of a meteorite impact on Earth had already been stated as a possible reason for the dinosaurs' extinction. However, after the 65 million-year-old rock layer iridium deposits were discovered, scientists had provable facts that a meteorite had struck the Earth at the same time as the mass dinosaur extinction. Geologists also found evidence of worldwide forest fires in the

65 million-year-old rock layer. Scientists used this information to propose a theory that the meteorite impact had caused massive worldwide fires that filled the air with soot and blocked much of the sunlight. As a result, the plant life and climate that dinosaurs ultimately depended upon for survival vanished and the mass extinction of the dinosaurs occurred. Proof of the validity of this theory was obtained when scientists discovered an extremely large 65 million-year-old meteorite crater near the Yucatan peninsula in Mexico.

A troubling scientific mystery concerning the reason why dinosaurs became extinct was solved. There was never any evidence found to support the other opinions regarding the dinosaurs' extinction, therefore, they were invalidated. The opinion which was based upon a meteorite impact was validated by the evidence cited above and was accepted as fact by the scientific community.

Questions

1. Write the opinions that were given for the massive extinction of the dinosaurs.

2. Why is the meteorite opinion now considered to be a fact? Explain.

3. Is the meteorite theory supported by provable evidence? Explain.

4. Which of the opinions held before the discovery of the Yucatan meteorite would you have supported and why?

Answers

1. The opinions given for the massive extinction of the dinosaurs included massive volcanic eruptions, radical climate change, a meteorite impact on Earth, or the eating of dinosaurs' eggs by mammals.

2. Evidence has been found that a large meteorite struck the earth at the time of the dinosaurs' extinction.

3. Yes, scientists can prove, based upon the amount of iridium and the existence of a large crater, that a meteorite struck the earth.

4. This answer will vary from student to student.

Questions

1. Circle the definition that best describes a FACT.

    A. A personal assumption based on observations

    B. Something that an observer believes to be true

    C. Something that is provable and widely accepted as being true

2. Circle the definition that best describes an OPINION.

   A. A personal belief that is based upon personal observations and assumptions, that cannot be proven through experimental means.

   B. A proven scientific theory.

   C. A belief that is based upon observable evidence and experimental evidence that is widely accepted by the scientific community.

3. What must occur for an idea that is considered to be an opinion to be accepted as a fact?

   _____
   _____
   _____

4. If a world-renowned scientist who is an expert in his or her field of study proposed a theory, would his status and position be enough to validate the theory as a fact?

   _____
   _____
   _____

Answers

1. (C) Choice (A) is incorrect because an opinion, not a fact, is based on personal assumptions and observations. Choice (B) is incorrect, because in order for something to be a fact, most people, not just one observer, must believe it to be true.

2. (A) Choice (B) is incorrect because opinions cannot be proven through experimental means, and scientific theories are tested through experiments. Choice (C) is incorrect for the same reason and also because an opinion can be held by a single person and not a scientific community.

3. Experimental evidence must be discovered to support the idea. The idea must be accepted widely by the scientific community.

4. No, the theory would have to be proven by experimental evidence and would have to be accepted widely by the scientific community.

UNDERSTANDING A PHOTO

Photographs can contain a lot of information about something. You can use your past experience to interpret a photograph along with any details supplied with it. For example, let's look at the photo shown below.

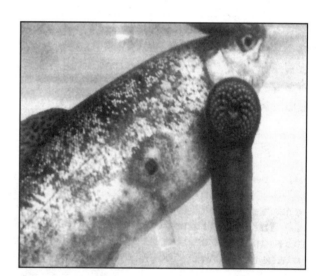

Questions

1. The fish in the photo has another animal attached to it just below its gills. This other animal is called a *lamprey*. The lamprey has a long, tubular body and a circular sucking disc around its mouth. The lamprey attaches itself to fish by means of this disc and uses its sharp teeth to feed on the blood and soft tissue of its host.

Do you see what is probably evidence of a lamprey wound on the unlucky fish in the photo? Write your observation in the space that follows.

2. Words above or below a picture explaining what it is about is called its caption. Write a caption for the photo on the previous page.

Answers

1. An example answer would be: "A round mark with a black center."

2. An example answer would be: "A fish that was wounded by a lamprey"

Whether we speak of animals that live on land, in the air, or underwater, it is possible to put them into categories known as species. All animals belong to the same species if they have: (a) similar appearances, (b) similar habits, (c) the ability to mate successfully, and (d) the ability to produce offspring which can also mate successfully, in a natural environment. Dogs and cats are examples of two different species; they cannot breed with one another. However, two closely related species may interbreed, as in the case of a horse and a donkey which mate to produce an offspring called a mule.

The mule itself is unable to reproduce further, which shows that the horse and the donkey are truly two different species.

Questions

Can you identify the animal shown below?

1. Write a short description of this animal.

This is a picture of a *duck-billed platypus*. It lays eggs and is found in Australia. Laying eggs makes it quite different from the kangaroo and the lemur, which both nurture their young inside their bodies.

2. What kind of surroundings do you think the platypus lives in by observing its webbed feet?

Answers

1. An animal with webbed feet, a bill, round back, fur, dark color, and tail.

2. The platypus probably lives in marshes or water.

TOPICS IN LIFE SCIENCE

This material will give you an overview of the life sciences field. You will need to become familiar with these topics before taking the GED examination.

Biology

Biology is the study of living things. Living things are differentiated from nonliving things by the ability to perform all the following life activities at some point in a normal life span.

Life Activity	Function
food getting	procurement of food through eating, absorption, or photosynthesis
respiration	exchange of gases
excretion	elimination of wastes
growth and repair	increase in size over part or all of a life span, repair of damaged tissue
movement	willful movement of a portion of a living thing's body, or direction of growth in a particular direction
response	reaction to events or things in the environment
secretion	production and distribution of chemicals that aid digestion, growth, metabolism, etc.
reproduction	the making of new living things similar to the parent organism(s)

It is important to note that living things *must*, during a typical life span, be able to perform all these activities. It is quite common for nonliving things to perform one or more of these activities, for example: robots—movement, response, repair; crystals—growth.

Cells

Cells are the basic structure unit of living things. A cell is the smallest portion of a living thing that can, by itself, be considered living. Plant cells and animal cells, though generally similar, are distinctly different because of the unique plant structures, cell walls, and chloroplasts. (See Figure 12)

Cells are made of several smaller structures, called organelles, that are surrounded by cell fluid, or cytoplasm. The functions of several cell structures are listed below.

Cell Structures	Function
cell membrane	controls movement of materials into and out of cells
cell wall	gives rigid structure to plant cells
chloroplast	contains chlorophyll, which enables green plants to make their own food
cytoplasm	jelly-like substance inside of a cell
mitochondria	liberate energy from glucose in cells for use in cellular activities
nucleus	directs cell activities; holds DNA (genetic material)
ribosome	makes proteins from amino acids
vacuole	stores materials in a cell

There are several processes cells perform to maintain essential life activities. Several of these processes, related to cell metabolism, are described on the next page. Metabolism is the sum of chemical processes in living things.

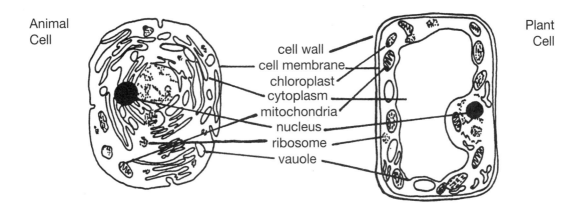

Figure 12: The Animal and Plant Cells

Process	Organelle	Life Activity
diffusion	cell membrane	food getting, respiration, excretion
osmosis	cell membrane	food getting, excretion
phagocytosis	cell membrane	food getting
photosynthesis	chloroplasts	food getting
respiration (aerobic)	mitochondria	provides energy
fermentation	mitochondria	provides energy

Cells need to move materials into their structures to get energy and to grow. The cell membrane allows certain small molecules to flow freely across it. This flow of chemicals from areas of high concentration to areas of low concentration is called diffusion. Osmosis is diffusion of water across a semipermeable membrane. Particles too large to be passed through the cell membrane may be engulfed by the cell membrane and stored in vacuoles until they can be digested. This engulfing process is called phagocytosis.

All cells need energy to survive. Sunlight energy can be made biologically available by converting it to chemical energy during photosynthesis. Photosynthesis is carried out in the chloroplasts of green cells. Chlorophyll, the pigment found in chloroplasts, catalyzes (causes or accelerates) the photosynthesis reaction that turns carbon dioxide and water into glucose (sugar) and oxygen.

$$6CO_2 + 6H_2O \xrightarrow[\text{chlorophyll}]{\text{sunlight}} C_6H_{12}O_6 + 6O_2$$

Sunlight and chlorophyll are needed for the reaction to occur. Chlorophyll, because it is a catalyst, is not consumed in the reaction and may be used repeatedly.

The term "respiration" has two distinct meanings in the field of biology. Respiration, the life activity, is the exchange of gases in living things. Respiration, the metabolic process, is the release of energy from sugars for use in life activities. Respiration, the metabolic process, occurs on the cellular level only. Respiration, the life activity, may occur at the cell, tissue, organ, or system level, depending on the complexity of the organism involved.

All living things get their energy from the digestion (respiration) of glucose (sugar). Respiration may occur with oxygen (aerobic respiration) or without oxygen (anaerobic respiration or fermentation). When respiration is referred to, it generally means aerobic respiration.

aerobic respiration:

$$C_6H_{12}O_6 + 6O_2 \rightarrow 6CO_2 + 6H_2O + energy$$

fermentation:

$$C_6H_{12}O_6 \rightarrow CO_2 + alcohol + energy$$

Aerobic respiration occurs in most plant and animal cells. Fermentation occurs in yeast cells and other cells in the absence of oxygen. Fermentation by yeast produces the alcohol in alcoholic beverages and the gases that make yeast-raised breads light and fluffy.

Classification

All known living things are grouped in categories according to shared physical traits. The process of grouping organisms is called classification. Carl Linné, also known as Linnaeus, devised the classification system used in biology today. In the Linnaeus system, all organisms are given a two-word name (binomial). The name given consists of a genus (e.g. Juniperus) and a species (e.g. communis) designation. Genus designations are always capitalized and occur first in the binomial. Species designations usually start with a lowercase letter and occur second. Binomials are usually underlined or italicized, e.g. *Genus species,* or *Homo sapiens,* or *Juniperus communis*.

There exists just one binomial for each organism throughout the scientific community. Similar genera of organisms are grouped into families. Families are grouped into orders, orders are grouped into classes, classes are grouped into phyla, and phyla are grouped into kingdoms. The seven basic levels of classification, listed from the largest groupings to the smallest, are: kingdom, phylum, class, order, family, genus, species.

Most biologists recognize five biological kingdoms today: *Animalia* (animals); *Plantae* (plants); *Fungi* (fungus); *Protista* (protists); and *Monera* (Monerans). Most living things are classified as plants or animals.

Monerans (e.g. bacteria, blue-green algae) are the simplest life forms known. They consist of single-celled organisms without a membrane-bound cell nucleus. Blue-green algae make their own food by photosynthesis; bacteria are consumers or parasites.

Protists (e.g. protozoa, single-celled algae) are single-celled organisms having cell nuclei. Protozoa (e.g. amoeba, paramecia) are predators or decomposers. Algae (e.g. Euglena, diatoms) are producers and utilize photosynthesis.

Fungi (e.g. molds, yeast) are many-celled decomposers that reproduce through spores. Yeasts are single-celled fungi that reproduce through budding. Fungi are the only many-celled decomposers that are not mobile.

Plants

Plants are multicellular organisms that make their own food through photosynthesis. Plants are divided into two phyla, the Bryophyta and Tracheophyta. Bryophytes are nonvascular plants. They lack true roots and woody tissues. Bryophyta (e.g. moss, liverworts, and multicellular algae) live in water or

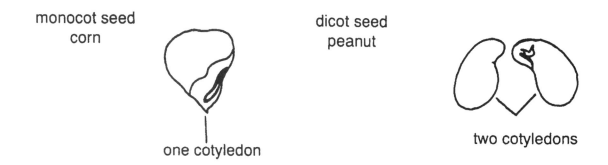

Figure 13

in damp areas and reproduce by spores. Bryophytes do not grow very tall because they lack the structural support of vascular tissue.

Tracheophytes are vascular plants. They have woody tissues and roots. The woody tissues in vascular plants enable them to grow quite large. The roots of vascular plants enable them to find water even in soils that are dry at the surface.

Tracheophytes are divided into three classes: *Filicinae*, *Gymnospermae*, and *Angiospermae*. Filicinae are ferns. They reproduce by spores. Gymnosperms (e.g. spruce, pines) are plants whose seeds form in cones. The seeds are unprotected. Angiosperms (e.g. apple trees, grass) are plants whose seeds are protected by fruits or other structures. Angiosperms are further divided into monocots or dicots, based on seed structure. Cotyledons are food storage structures in seed embryos. Monocots (e.g. grasses, bananas) have one cotyledon per seed. Dicots (e.g. oak trees, pumpkins) have two cotyledons per seed. (See Figure 13)

Animals

Animals are multicellular organisms that cannot make their own food but can move themselves about. The animal kingdom is split into 26 phyla. Some of the phyla are listed on the top of the following page.

The arthropod and chordate phyla deserve special note. The arthropods include 10 classes, three of which are very important to humans: the *Arachnidae*, *Insecta*, and *Crustaceae*. Arachnids include spiders and ticks. These animals have two body regions and eight legs. Insects have three body regions and six legs. They include an incredible variety of animals; for example, grasshoppers, flies, beetles, and butterflies are typical insects. Crustaceans have two body regions and ten legs and live mostly in water. Crabs, crayfish, and lobsters are all crustaceans.

The chordate phylum has three subphyla, one of which is the vertebrata, or vertebrates. Vertebrates have an internal skeleton that includes a spine made up of vertebrae. The spine protects the dorsal nerve cord (spinal cord). Animals without spines (e.g. all phyla except Chordata) are called invertebrates.

Vertebrates

Eight classes of vertebrates exist, though four are often spoken of collectively as "fish."

Examples of Animal Phyla

Phyla	Examples	Traits
Porifera	sponges	no organs, pores in body let water flow through, bringing food and taking away wastes
Coelenterata (Cridaria)	jellyfish, corals, hydra	no organs, body sac-like, stinging cells to capture prey
Platyhelminthes	flatworms, flukes, tapeworms	single opening to body, true organs, often parasitic to humans
Aschelminthes (Nematoda)	roundworms	two openings to body, may be parasitic
Mollusca	snails, octopi, clams	gills, open circulatory system, produce shells (internal or external)
Annelida	earthworms, leeches	closed circulatory system
Arthropoda	spiders, insects, crabs	jointed exoskeletons, jointed legs
Echinodermata	starfish, sea urchins	plate-like internal skeleton, tube feet, spiny or knobby surface
Chordata	fish, birds, mammals, reptiles	notochord (primitive cartilaginous spine), pharyngeal gill slits, and hollow dorsal nerve cord are present at some point in development

Classes of Vertebrates

Class	Traits	Examples
Agnatha (FISH)	jawless fish, no scales, cartilaginous skeleton	lampreys, hagfish
Placodermi (FISH)	hinged jaws	extinct
Chondrichthyes (FISH)	cartilaginous skeleton, no scales, jaws	sharks, skates, rays
Osteichthyes (FISH)	bony skeleton, scales, jaws	bass, trout, goldfish
Amphibia	aquatic eggs and larvae, terrestrial adults	frogs, toads, salamanders
Reptilia	terrestrial eggs and adults, cold-blooded	turtles, snakes, lizards
Aves	feathers, warm-blooded, external egg development	eagles, ducks, pigeons
Mammalia	fur, milk-producing, internal egg development, warm-blooded	rats, horses, humans

Most vertebrates are cold-blooded. Their bodies do not generate heat, so their body temperature is determined by their environment. Fish are cold-blooded animals with gills for respiration and fins for limbs. Reptiles are cold-blooded animals with lungs for respiration and legs for limbs (except for snakes). Amphibians are cold-blooded animals that start life with gills and fins but then change. The change in form that amphibians undergo as they mature is called metamorphosis. Adult amphibians have lungs and legs.

Birds (*Aves*) and mammals are warm-blooded. Their bodies generate heat. Birds and mammals can also sweat to lower body temperature. Birds are covered with feathers and have eggs that develop outside the mother's body. Mammals are covered with fur and have eggs that develop within the mother's body.

Mammals are divided into 17 orders, based on body structure. Some of the more familiar orders are listed below.

Viruses

Viruses are organic particles that are capable of causing diseases in living things, such as smallpox, rabies, and influenza. Viruses are sometimes classified as living things because they contain genetic material and create offspring similar to themselves. Viruses are often not classified as living things because they have no ability to synthesize or process food and cannot reproduce without the help of other organisms. Viruses are parasitic. Their basic structure is a protein shell surrounding a nucleic acid core.

Ecology

Ecology is the study of the relationship between living things and their environment. An environment is all the living and nonliving things surrounding an organism.

Populations and Communities

A population is a group of similar organisms, like a herd of deer. A community is a group of populations that interact with one another. A pond community, for example, is made of all the plants and animals in the pond. An ecosystem is a group of populations which share a common pool of resources and a common physical/geographical area. A beech-oak-hickory forest ecosystem, for example, is made of populations in the forest canopy, on the forest floor, and in the forest soil.

Each population lives in a particular area and serves a special role in the community. This combination of defined role and living

Familiar Order of Mammals

Order	Examples	Traits
Marsupials	kangaroos, opossums	pouches in mothers for carrying young
Rodents	mice, rats, beavers, squirrels	gnawing teeth
Carnivores	dogs, bears, cats, skunks	meat eaters
Cetaceans	whales, dolphins, porpoises	aquatic, flippers for limbs
Primates	monkeys, apes, humans	opposable thumbs, erect posture, highly developed brains
Ungulates (2 orders)	horses, camels, buffaloes	grass chewers

areas is the concept of niche. The niche of a pond snail, for example, is to decompose materials in ponds. The niche of a field mouse is to eat seeds in fields. When two populations try to fill the same niche, competition occurs. If one population replaces another in a niche, succession occurs. Succession is the orderly and predictable change of communities as a result of population replacement in niches.

A climax community is a community in which succession no longer occurs. Climax communities are stable until catastrophic changes, such as forest fires, hurricanes, or human clearing of land occurs. Each ecosystem type is defined by its climax communities, for example, beech-oak-hickory forests in the American Northeast or prairies in the American Midwest.

Food and Energy

Energy enters ecosystems through sunlight. Green plants turn this energy into food in the process of photosynthesis. Organisms that make their own food are called producers. Some animals get their food from eating plants or other animals. Animals that get their food energy from other living things are called consumers. Consumers that eat plants are herbivores; those that eat animals are carnivores; those that eat plants and animals are called omnivores. Animals that eat other organisms are called predators; the organisms that get eaten are called prey. Organisms that get their food energy from dead plants or animals are called decomposers.

As energy moves from one organism to another, it creates a pattern of energy transfer known as a food web. (See Figure 14)

Arrows represent energy transfer in a food web. At each energy transfer (arrow) some energy is lost. Energy is lost because organisms use energy to grow, move, and live.

Many nutrients, such as nitrogen and phosphorous, are routinely cycled through the bodies of living things. These nutrient cycles are disrupted when humans remove parts of the ecosystem or add excess materials to an ecosystem.

Climate

Climate has a direct effect upon humans and other living organisms. When referring to climate, one is talking about the usual weather that occurs in a general area over a period of time. It takes into consideration air temperature, wind speed, sunshine, humidity, amount of precipitation, air pressure, and general geographic conditions. Climate has a direct influence on people and other living organisms. For example, it affects the availability of food and water. It also influences people's method of transportation, outdoor activities, choices for employment, type of clothing, and type of housing.

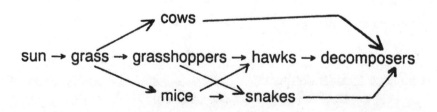

Figure 14: Example of a Food Web

Pollution and Conservation

Pollution is damaging to both the ecosystems and living organisms. Any material added to an ecosystem that disrupts its normal functioning is pollution. It affects air, water, soil, and food resources. People are contributing greatly to Earth's pollution by producing an overwhelming amount of waste. This includes exhaust fumes from cars, household and industrial wastes, fertilizers and pesticides, and radioactive wastes from power plants. The growing population and modern conveniences accelerate pollution. People have to change their habits in order to solve this problem and ensure their survival.

Conservation is the protection and management of Earth's natural resources, for the benefit of not only man, but all living things. Without wise practices of conservation and concern for the quality of the environment, all natural resources necessary for life, such as air, animals, energy, minerals, plants, soil, and water would be damaged, wasted, or destroyed. Earth's resources are limited and with demands growing as populations increase they have to be managed wisely. Recycling plastic, glas, and paper is one way to save resources and help conservation.

Human Health

The Human Body

A human is a very complex organism. This complexity requires individual cells to become specialized at certain tasks. Groups of specialized cells form tissues, such as muscles, skin, or blood. Tissues are specialized to perform specific tasks. Groups of tissues form organs, such as the heart, kidney, or brain. Systems are groups of organs working together to perform the basic activities of life, such as excretion or reproduction. The human body has several systems in it.

The skeleton supports the body and gives it shape. The skeletal system is composed of bones, cartilage, and ligaments. The human body has 206 bones in it. The areas where two or more bones touch one another are called joints. Five types of joints exist in the human body: fixed joints (skull bones), hinge joints (elbow or knee), pivot joints (neck bones), sliding joints (wrist bones), and ball-and-socket joints (shoulder or hip). Bone surfaces in joints are often covered with cartilage, which reduces friction in the joint. Ligaments hold bones together in a joint.

The muscular system controls movement of the skeleton and movement within organs. Three types of muscle exist: striated (voluntary), smooth (involuntary), and cardiac. Cardiac muscle is found only in the heart and is involuntary. Smooth muscle is found in organs and cannot be consciously controlled (involuntary). Striated muscle is attached to a skeleton and its actions can be controlled at will, or voluntarily. Tendons attach muscles to bone. Muscles perform work by contracting. Skeletal muscles work in pairs. The alternate contraction of muscles within a pair causes movement in joints.

The digestive system receives and processes food. The digestive system includes the mouth, stomach, large intestine, and small intestine. Food is physically broken down by mastication, or chewing. Food is chemically broken down in the stomach, where digestive enzymes break the food down into simple chemicals. The small intestine absorbs nutrients from food. The large intestine absorbs water from solid food waste.

The excretory system eliminates wastes from the body. Excretory organs include the lungs, kidneys, bladder, large intestines, rectum, and skin. The kidneys filter blood and excrete wastes, mostly in the form of urea. The bladder holds liquid wastes until they can be eliminated via the urethra. The large intestine absorbs water from solid food waste, and the rectum stores solid waste until it can be eliminated. The skin excretes waste through perspiration. The lungs excrete gaseous waste.

The circulatory system is responsible for internal transport in the body. It is composed of the heart, blood vessels, lymph vessels, blood, and lymph. The heart is a muscular four-chambered pump. The upper chambers are called atria and the lower chambers are called ventricles. Blood flows from the body to the (1) right atrium, to the (2) right ventricle, to the lungs, then to the (3) left atrium, to the (4) left ventricle, and back to the body. (See Figure 15, The Heart)

The heart chambers contract to expel the blood they contain. Blood flows in one direction through the heart because of valves within the heart and blood vessels. The closing of valves during heart contractions creates an audible heartbeat. An adult human heart normally contracts 60–80 times per minute.

There are three types of blood vessels: arteries, veins, and capillaries. Arteries have thick, muscular walls and carry oxygenated blood away from the heart. Veins have thin walls and carry deoxygenated blood to the heart. Capillaries have extremely thin walls and connect arteries to veins.

Blood is always under pressure in the arteries. Blood pressure increases when the heart is contracting. Blood pressure during heart contractions is systolic pressure. Blood pressure during heart relaxation is diastolic pressure. Human blood pressure is always reported as a ratio of systolic pressure/diastolic pressure. Typical blood pressure for adults is 140 mm Hg/90 mm Hg. Pressures ranging far above or below these values indicate illness.

The fluid portion of blood is called plasma. The solid material in blood includes red blood cells, white blood cells, and platelets. Red blood cells carry oxygen to cells and carry carbon dioxide away from cells. White blood cells fight infections and produce antibodies. Platelets cause the formation of clots.

The lymphatic system drains fluid from tissues. Lymph nodes filter impurities in the lymph fluid, and often become swollen during infections.

The respiratory system exchanges oxygen for carbon dioxide. The respiratory system is composed of the nose, trachea, bronchi, lungs, and diaphragm. Air travels from the nose through the trachea and bronchi into the lungs. The air is drawn in by the contraction of the diaphragm, a muscle running across the body below the lungs. Gas exchange occurs in the lungs across air sacks called alveoli. Air is then pushed back toward the nose by relaxation of the diaphragm.

The nervous system controls the actions and processes of the body. The nervous system includes the brain, spinal cord, and nerves.

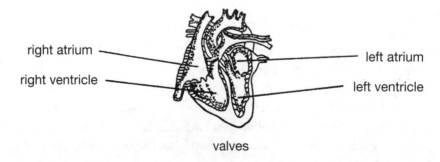

Figure 15: The Heart

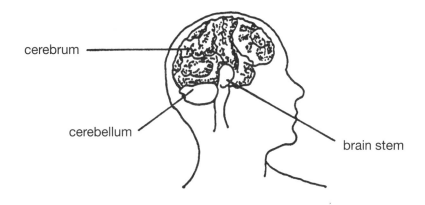

Figure 16: The Brain

Electrical impulses carry messages to and from the brain across the spinal cord and nerves. Nerves extend to every portion of the body. The spinal cord is protected by the backbone.

The three principal regions of the brain are the cerebrum, cerebellum, and brain stem. The cerebrum occupies 80% of the brain's volume and is responsible for intelligence, memory, and thought. The cerebellum is located at the lower rear portion of the brain, and it controls balance and coordination. The brain stem connects the brain to the spinal cord and is found at the lower central portion of the brain. The brain stem controls autonomic (involuntary) body functions and regulates hormones. (See Figure 16)

The endocrine system controls activities in the body through chemical agents called hormones. Hormones are produced in glands throughout the body and are excreted into the bloodstream. The brain controls production and release of hormones.

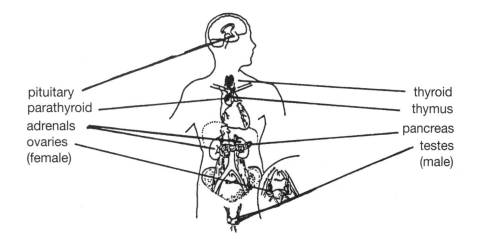

Figure 17: Major Endocrine Glands

Gland	Hormone	Action
hypothalamus	oxytocin	stimulates labor in childbirth and production of milk in females
pituitary	growth hormone	stimulates growth
thyroid	thyroxin	controls rate of cellular respiration
parathyroid	parathormone	controls amount of calcium in the blood
thymus	thymosin	helps to fight infections
adrenals	adrenalin (epinephrine)	helps during stress and shock and activates flight-or-fight response
pancreas	insulin	regulates blood sugar
ovaries (female)	estrogen, progesterone	controls female maturing process, maintains pregnancy
testes (male)	testosterone	controls male maturing process

Listed above are several major endocrine glands and the most important hormones they produce. (See Figure 17)

The reproductive system produces eggs (females) and sperm (males) which can combine to create an embryo. The female reproductive system includes the ovaries, fallopian tubes, uterus, and vagina. Each month one egg is released from one of the ovaries and then travels down the fallopian tubes. If it is fertilized, then it becomes implanted in the lining of the uterus, where an embryo begins to form. When sufficiently grown, the fetus leaves the uterus and its mother's body through the vagina, or birth canal.

The male reproductive system consists of the testicles, vas deferens, urethra, and penis. Sperm are produced in the testicles. They move through the vas deferens from the testicles to the urethra. During unprotected intercourse, sperm pass through the penis (via the urethra) and into the woman's body. In the woman's body, sperm pass through the cervix into the uterus and up the fallopian tubes, where fertilization of an egg may take place.

Disease and Prevention

Bacteria and viruses cause diseases like tuberculosis, pneumonia, food poisoning, or the flu. The human body possesses many defenses against parasites, among them the skin, certain white blood cells called phagocytes, and disease-fighting cells called antibodies. Diseases can spread in various ways: through sneezing and coughing, through contaminated water and food, or through contact with an infected person or animal. Some ways of controlling the spread of diseases include disinfection, sterilization, and personal cleanliness. Today, many vaccines are available to prevent infection in the first place. However, scientists still try to find a way to battle serious illnesses like cancer or AIDS.

Vitamins Important to Human Health

Vitamin	Principal Source	Deficiency Symptom(s)
A	green and yellow vegetables	night blindness, dry, brittle skin
B_1	cereals, yeast	beriberi (muscular atrophy and paralysis)
B_2	dairy products, eggs	eye problems
B_{12}	liver and meat	anemia
C	citrus fruits, tomatoes	scurvy
D	fortified milk, eggs	rickets (malformed bones)
E	meat, oils, vegetables	male sterility, muscular problems
K	green vegetables	impaired blood clotting

Nutrition

Nutrition is the study of how living things utilize food substances. Food provides energy and raw materials for growth, repair, and metabolism. Energy (calories) is derived chiefly from carbohydrates, but also from fats and proteins. Raw materials for life processes come chiefly from protein, but also from carbohydrates (starches and sugars), fats, minerals, and vitamins. Fiber in the diet helps in elimination of wastes.

Listed above are several vitamins important to human health.

Foods can be placed in one of four basic food groups. A healthy diet is one that includes food from each of these groups daily, as shown below.

Human Genetics

Each of the cells in a living thing has a specific structure and role in the organism. The structure of a cell and its function are determined, to a large degree, by the genes within a cell. Genes are code units of chromosomes within the nucleus of a cell. Genes give information about the structure and function of a cell.

Cells age and die. Organisms continue to live despite the death of individual cells because cells reproduce. Mitosis is the process of cell reproduction through cell division; one cell divides to become two new cells. During mitosis, the genes in the parent cell are copied and passed to the offspring. Thus, each new cell contains the same genetic material as the parent cell. The passing of genetic material from

Food Group	Importance	Examples
Meat, Fish, and Eggs	protein	steak, trout
Dairy	fats, calcium, protein	milk, cheese
Fruits and Vegetables	fiber, minerals, vitamins	apples, carrots
Grains and Cereals	starch (for energy), protein, fiber	bread, pasta

one generation to the next is called inheritance. The study of genetic material and inheritance is called genetics.

Most cells in the human body have 46 chromosomes. Some special cells, called eggs and sperm, have only 23 chromosomes. Egg and sperm cells get 23 chromosomes through meiosis, a process of cell division that reduces the number of chromosomes in a cell. Most cells in the human body reproduce quite often. Egg and sperm cells (sex cells) cannot reproduce until they join with another sex cell. The process of an egg and sperm cell joining is called fertilization.

A fertilized human sex cell has 46 chromosomes, 23 from the mother and 23 from the father. This fertilized sex cell will multiply to form a new organism. The genes of the new organism are a mixture of genes from both parents, so the new organism will be unique from each parent. The process of combining genetic materials from two parent organisms to form a unique offspring is called sexual reproduction.

During sexual reproduction an organism receives two genes for each trait, one from each parent. Sometimes one trait will mask another, as is the case with eye color. If a person has one gene for brown eyes and one gene for blue eyes, the person will always have brown eyes. A genetic trait that masks another, like the gene for brown eyes, is called a dominant trait. A gene that can be masked, like the gene for blue eyes, is called a recessive trait.

Understanding dominance helps us to figure out the genetic configuration of an individual. An individual with blue eyes must have two genes for blue eyes, since it is a recessive trait. Recessive traits are shown by lowercase letters, so the genetic symbol for blue eyes is "bb." An individual with brown eyes must have at least one gene for brown eyes, which is dominant. Dominant genes are shown by uppercase letters, so the genetic symbol for brown eyes could be "Bb" or "BB."

An individual with two different genes (e.g. Bb) for a trait is called heterozygous for that trait. An individual with two similar genes (e.g. BB or bb) for a trait is called homozygous for that trait.

When the genetic type of parents is known, the probability of the offspring showing particular traits can be predicted using the Punnett square. A Punnett square is a large square divided into four small boxes. The genetic symbol of each parent for a particular trait is written alongside the square, one parent along the top and one parent along the left side.

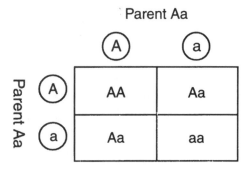

Figure 18: Punnett square

Each gene symbol is written in both boxes below or to the right of it. This results in each box having two gene symbols in it. The genetic symbols in the boxes are all the possible genetic combinations for a particular trait of the offspring of these parents. Each box has a 25% probability of being the actual genetic representation for a given child.

The genetic symbol "AA" in the example has a 25% probability of occurring. If a genetic symbol occurs in more than one box, the genetic probabilities are added. The genetic code "Aa" in the example has a 50% probability of occurring because it is shown in two boxes of the Punnett square.

The Punnett square shows how two parents can have a child with traits different from either parent. Two parents heterozygous for brown eyes (Bb), have a 25% probability of producing a child homozygous for blue eyes (bb), as shown in Figure 18.

Human sex type is determined by genetic material in sperm. The genetic sex code for human females is XX. The genetic sex code for human males is XY. Eggs carry only X genes. Sperm carry X or Y genes. The probability of a fertilized human egg being male, or XY, is 50%.

Human blood type is determined genetically. Genes for blood type may be one of three kinds, i, I^A, or I^B. The i gene is recessive. Blood type O is caused by an ii genetic code. The I^A and I^B genes are dominant. They may be represented by $I^A i$ or $I^A I^A$ and $I^B i$ or $I^B I^B$, respectively. Blood type AB has genetic code $I^A I^B$. Blood types fit the Punnett square model.

Genes and the Environment

An organism's particular genetic makeup will determine its physical characteristics. The expression of some genes is sometimes dependent upon the environmental conditions, for example, temperature or availability of light. Differences in the appearance or intelligence of identical twins that were separated at birth demonstrate the importance of environment in gene expression.

☞ Practice: Life Sciences

DIRECTIONS: Fill in the blanks.

Questions 1–6 are based on the following passage.

Humans are unique in their ability to modify the carrying capacity of their environment to be favorable for population growth. Most scientists agree that it is only a matter of time before humans will be unable to increase the Earth's carrying capacity any further. At that time, the birth rate must decrease and the death rate must increase to balance population growth.

Natural resources are necessary for human survival and the making of necessary products. The natural resources are water, soil, air, wildlife, and forests. Problems that are now being faced are related to erosion, soil depletion, species extinction, deforestation, desertification, and water shortages. Efforts to reverse these problems and their environmental damages are found in the planned programs of reforestation, captive breeding, biological harvesting, or planned farming through efficient plowing and planting procedures.

Pollution is damaging both the ecosystems and living organisms. Air, water, soil, and food resources are being affected by pollution. Pollutants include automobile exhaust, fertilizers, pesticides, industrial wastes, radioactive wastes, and most of all, household wastes. The growing population and modern conveniences greatly contribute to this insurmountable problem. Government regulations, community efforts, and changes in the habits of industries and individuals are necessary to solve pollution problems.

1. Based on the passage, what are some of the effects of damage of natural resources by human beings?

   _____
   _____
   _____

2. What are some of the causes of the depletion of our natural resources?

   _____
   _____
   _____

3. What do most scientists predict will happen if the growing population is not slowed down?

   _____
   _____
   _____

4. If one were to conclude from this passage that humans will exhaust the Earth's carrying capacity in the near future, would this be fact or opinion? Explain your answer.

   _____
   _____
   _____

5. What is the main idea of this passage?

   _____
   _____
   _____

6. Based on its use in this passage, what do you think the meaning of the word **desertification** is?

   _____
   _____
   _____

Questions 7–11 are based on the following diagram.

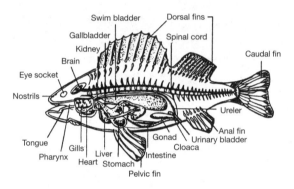

7. The fish's heart is closest to the middle of its body. T   F

8. Fish have five fins. T   F

9. The long pouch directly under the spinal cord is the liver. T   F

10. The spinal cord starts at the brain and ends at the caudal fin. T   F

11. The mouth cavity of the fish is called the tongue. T   F

Questions 12–15 are based on the following paragraph.

When you look at a picture, always try to see things which seem unusual compared to what you expect it to show. In the case of the fish and the lamprey on page 38, it seems unusual to have a long tubular object attached to the side of a fish! Look at the pictures on the following page.

Life Sciences

12. Write a short description of the picture on the left.

    _____

    _____

    _____

13. Write a short description of the picture on the right.

    _____

    _____

    _____

   Note that the picture on the left is a kangaroo; the picture on the right is a ring-tailed lemur.

14. What three similarities can you find between these two animals?

    _____

    _____

    _____

15. What are three differences between them?

    _____

    _____

    _____

Answers

1. Some of the problems with the environment that are related to human consumption are erosion, soil depletion, species extinction, deforestation, desertification, and water shortage. One could also cite from the last paragraph that pollution is damaging both the ecosystem and living organisms.

2. Some of the causes for the depletion of natural resources have been pollutants, that include automobile exhaust, fertilizers, pesticides, industrial wastes, radioactive wastes, and household wastes.

3. They are predicting that the Earth will eventually run out of room to hold all its inhabitants. Before this happens, though, it can be assumed that there will be many serious problems as the number of people begins to outgrow the availability of natural resources.

55

4. This would be an opinion. Although it is agreed by most scientists that if the population growth continues at the present rate the Earth will be above its carrying capacity, they do not know that this will be in the near future.

5. The main idea of this passage is that the Earth is faced with serious problems that threaten the Earth's natural resources and all its inhabitants and that it is up to human beings to alter the trend toward destruction and make efforts to preserve the Earth.

6. Based on its grouping with words such as soil depletion, extinction, and erosion we can assume that *desertification* is something bad. If we recognize the root of the word, *desert,* we can guess that it means the expansion of desert lands or the changing of good soil into desert land.

7. False. The fish's heart is closest to the front of its body.

8. True. Fish have five fins: two dorsal fins, one caudal fin, one pelvic fin, and one anal fin.

9. False. The long pouch directly under the spinal cord is the swim bladder, not the liver.

10. True. The spinal cord starts at the brain and ends at the caudal fin.

11. False. The mouth cavity of the fish is called the pharynx, not the tongue.

12. The picture shows a kangaroo with its baby in its front pouch. Notice its short front legs, large hind legs for jumping, and its long tail for balance.

13. It is a ring-tailed lemur sitting in a field.

14. Both the lemur and kangaroo have fur, tails, and a round back.

15. The animals' head shape and color differ. Also, the lemur is not carrying a baby in a front pouch.

REVIEW

In this section we learned some very valuable skills used in approaching material in life sciences. These skills are making inferences, finding the main idea, making predictions, distinguishing fact from opinion, summarizing information, recognizing cause and effect, and understanding diagrams and photographs. You will put all these skills to use in the following chemistry practice section.

Science

Chemistry

SCIENCE

CHEMISTRY

FINDING THE MAIN IDEA

To find the main idea of a scientific passage it helps to "pre-read" it by reading only the first sentence of each paragraph and then skimming the rest of it for key words. Try this method on the passage below and then answer the question which immediately follows it.

Chemistry is the science of matter and its changes at the atomic level. The basic building blocks of matter, called **atoms**, are the lowest level of structure studied in the field of chemistry. The changes that occur at the atomic level are called **chemical reactions**. Atoms of the same kind are said to be atoms of the same chemical element. The simplest atom and element is hydrogen. The hydrogen atom only has two parts. The next simplest element is helium. The helium atom has only six parts. About ten different kinds of atoms or elements are known. Many of the more recently discovered atoms or elements can only be created artificially in a laboratory.

Substances that are made of more than one element are called **compounds**. The smallest unit of a compound is called a **molecule** which in turn is made from atoms or elements. When a substance is produced by merely mixing one or more elements without changing the atomic bonds, it is called a **mixture**.

Chemical changes in matter are called **reactions**. Reactions obey natural laws which dictate how atoms make or break bonds to form new compounds.

Although chemical reactions have been observed since the taming of fire, progressive chemical studies began with the discovery of elements such as arsenic and antimony in the 1200s. The outlines of a completely correct and organized chemistry first appeared in the works of Antonio Lavoisier in 1788. He is often referred to as "the father of modern chemistry." Lavoisier corrected a previous error in thinking that had to do with burning. Before his time, people believed that fire released an element called phlogiston. Lavoisier corrected this by showing that it was really the absorption of oxygen that was taking place when burning occurred.

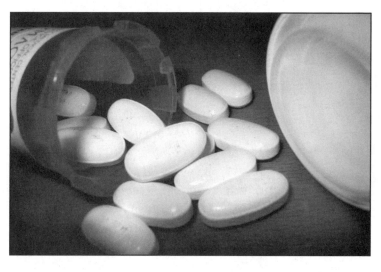

A second great chemistry error was corrected by Count Rutherford in 1820. He showed that heat was not a material substance, but was rather the result of atomic reactions. The correction of this error greatly strengthened the theories of modern chemistry.

What do you think the passage is about?

From reading only the first sentence of each paragraph you would know that the passage has something to do with chemistry and the different aspects of it. You may also notice such key words as *mixture, molecule,* or *reaction.*

Now go back and read the passage thoroughly and then answer the questions below to see how well you understand the material.

Questions

1. Circle the statement that represents the main idea of the preceding passage.

    (1) Chemistry has gone through revisions throughout the years.

    (2) The main concepts in chemistry, such as elements, compounds, and the nature of heat, form the foundations of chemistry.

    (3) Chemistry is unlike any other field of science because it has had revisions, and it has main concepts.

2. Circle the correct answer that completes the following sentence.

    The basic building blocks of matter and the lowest level of structure studied in chemistry are called

    (1) atoms.

    (2) mixtures.

    (3) compounds.

Circle "T" if the statement is true; circle "F" if the statement is false.

3. Modern chemistry began to develop in the 1200s.

    T   F

4. The simplest atom and element is helium.

    T   F

5. Chemistry has undergone revisions over time.

    T   F

Answers

1. (2); "Elements," "compound," and "heat," are all key words. Choices (1) and (3) are incorrect, because revisions are mentioned in only two of the five paragraphs of the passage. Therefore, they do not contain the main idea.

2. (1); Mixtures occur when a substance is produced by merely mixing one or more elements without changing the atomic bonds (2). Compounds are substances that are made of more than one element.

3. True. Although chemical reactions have been observed since the taming of fire, progressive chemical studies began with the discovery of antimony and arsenic in the 1200s.

4. False. The simplest atom and element is hydrogen, not helium. Hydrogen has two parts; helium has six parts.

5. True. Chemistry has undergone revisions over time, including errors corrected by Antonio Lavoisier and Count Rutherford. Before 1788, people believed that fire released phlogiston, an erroneous element. Lavoisier corrected this by showing that it was really the absorption of oxygen that was taking place when burning was occurring. In 1820, Rutherford showed heat was not a material substance

as believed, but was rather the result of atomic reactions.

COMPARING AND CONTRASTING

Comparing and contrasting are analytical skills that help us see the relationship between two things. It is very important when dealing with scientific material to be able to understand relationships. Being able to pinpoint the important details of a passage is important, but being able to take these details and see how they relate to, and affect, one another shows that you have a much deeper understanding of the material.

Recognizing relationships is a necessary tool even in day-to-day life. We employ this tool when we need to make a decision about a course of action. For instance, when we are making plans for a trip—let's say to the beach or the park—and we want to see which one is better, we will often make a mental list of all the benefits and all the drawbacks of each. We would then compare these items and see which is better.

In the example above, let's just say that the beach costs money. You may decide to go to the beach because it has a place to go swimming and you don't care about the money. You may, however, decide to go to the park because you would like to be outside in the sun but you do not want to pay for the beach. You *compared* the two choices and found that they both involve having a good time outside. You also *contrasted* the two places by recognizing that one is free and one is not, and that one has a place to swim and one doesn't.

Take a look at the recipe for cooking Old Fashioned Chocolate Fudge and then answer the questions that follow it to see how well you understand comparing and contrasting.

Old-Fashioned Chocolate Fudge

2 cups sugar

2/3 cup milk

2 ounces unsweetened chocolate or 1/3 cup cocoa

2 tablespoons corn syrup

1/4 tablespoon salt

2 tablespoons butter or margarine

1 teaspoon vanilla

1/2 cup chopped nuts (if desired)

Butter a loaf pan, 9 x 5 x 3 inches. Combine sugar, milk, chocolate, corn syrup, and salt in a 2 quart saucepan. Cook, stirring occasionally until the temperature of the mixture reaches 234 degrees Fahrenheit on a candy thermometer placed in the mixture (or until a very small amount of mixture dropped into very cold water forms a soft ball that flattens when removed from the water). Remove from the heat and add butter. Cool the mixture to 120 degrees Fahrenheit without stirring (bottom of the pan will be lukewarm). Add vanilla, beat vigorously and continuously for 5 to 10 minutes with a wooden spoon, until the mixture is thick and no longer glossy in appearance (mixture should hold its shape when dropped from a spoon). Quickly stir in nuts (if desired). Spread the mixture evenly in a buttered pan. Cool until firm. Cut into squares. One pound equals 32 one-inch squares.

Penuche variety: Substitute 1 cup of brown sugar for 1 cup of the granulated sugar and omit the chocolate or cocoa.

The cooking of fudge illustrates many chemical principles. In a typical type of fudge, sugar, milk, chocolate, corn syrup, butter, and salt are all stirred together and held over a cooking flame in a pot. The chemicals in the mixture react with oxygen in the air to begin their heat-induced chemical change. With ingredients fixed, as in the recipe for fudge, the cooking is done until the temperature of 234 degrees Fahrenheit is achieved on the candy thermometer and the mixture forms a soft ball in cold water that flattens when removed.

In a variant recipe, butter is added at the start of the cooling to 120 degrees Fahrenheit. The hot fudge is sensitive and is not stirred until it is cool. The cooler fudge (120 degrees Fahrenheit or less) is then beaten continuously for 5 to 10 minutes until the fudge is thick and not glossy. The mixture should hold its shape when dropped from a spoon. The fudge is then allowed to cool some more before serving.

The taste and texture of fudge is very sensitive to the exact quantities of ingredients included and to the exact heating and cooling procedures required by the fudge-making process. A common error is to cook fudge above 234 degrees Fahrenheit, which results in a grainy, almost purely sugar residue which does not taste very good. Also, letting the fudge cool beyond 120 degrees Fahrenheit before adding the vanilla can cause the vanilla to be unevenly distributed and negatively affect the taste of the fudge. Note that our taste buds and sense of smell also function due to chemical reactions and that an excess of one ingredient in the fudge would affect these senses differently.

Questions

1. What comparison can you make between fudge that is cooked above 234 degrees Fahrenheit and fudge in which the vanilla is added when the fudge has cooled below 120 degrees Fahrenheit?

2. How would fudge that is made properly contrast with fudge that is overcooked?

3. How would the penuche variation compare and contrast with traditional fudge?

Answers

1. In both cases, the taste of the fudge is negatively affected.

2. Fudge that is overcooked has a grainy residue that does not taste good. Fudge that is made properly is not grainy and tastes good.

3. The textures of the two types would be similar. The taste of the fudge would be different.

ROOTS, SUFFIXES, AND PREFIXES

The chemical formula or the name of a given chemical compound tells us the elements or compounds that make up a larger compound. Chemical compounds that have only hydrogen and carbon with only single bonds are called alkanes. The chemical names for alkanes all end in the suffix "ane." Alkanes also have prefixes in their chemical names that indicate the number of carbon atoms appearing in the compound. The first four prefixes are unusual, but after that the traditional Latin word roots are used.

Number of Carbon Atoms	Prefix	Name of Compound
1	meth–	methane
2	eth–	ethane
3	prop–	propane
4	but–	butane
5	pent–	pentane
6	hex–	hexane
7	hept–	heptane
8	oct–	octane
9	non–	nonane
10	dec–	decane

Carefully examine the chart above, which clearly demonstrates how to add the prefix to an alkane name.

Questions

1. What elements are present in decane?

2. How many carbon atoms are present in decane?

3. How many carbon atoms are present in butane?

4. An alkane with two carbon atoms is called _____.

Answers

1. Hydrogen and carbon are present in decane. Chemical compounds that have only hydrogen and carbon with only single bonds are called alkanes. The chemical names for alkanes all end in the suffix "-ane." Decane ends in the suffix "-ane." Therefore, decane is an alkane and only has hydrogen and carbon present in the compound.

2. 10; There are 10 carbon atoms present in decane. Alkanes have prefixes in their chemical names that indicate the number of carbon atoms appearing in the compound. The prefix for decane is "dec-." If you examine the chart, you will see that there are 10 carbon atoms in decane, and the prefix for "dec-" stands for 10.

3. 4; The prefix for butane is "but-." If you examine the chart, you will see that there are four carbon atoms in the compound butane, and the prefix "but-" indicates four.

4. The alkane with two carbon atoms is ethane. When you examine the chart, you will find the row with "number of carbon atoms." Find "2." The prefix for two is "eth-." The name of the compound is ethane.

MAKING PREDICTIONS

An important practice for scientists, particularly in the field of chemistry, is making predictions. This is an important skill because we get to apply our knowledge to a situation. This is also an important real-life skill because it helps us see how some things turn out based on what we already know about the situation in which we find ourselves.

Read the following passage and then answer the questions that follow it.

> Burning, also called combustion, occurs when a chemical compound combines with oxygen from the air through the aid of a flame. The heat of the flame triggers a chemical reaction between the chemical compound and oxygen that feeds upon itself and continues until the reaction is broken.
>
> A classic example is the wax candle. The wick, a string in the middle of the candle, is lit with a match. The candle continues to burn as a chemical reaction between the wax, and the oxygen in the surrounding air continues the burning process. We can observe that as long as the conditions for burning remain favorable, a candle will burn until the wick and wax are consumed.

Questions

1. Give a definition of "burning."

2. What factors could cause the burning process of a candle to stop?

3. When a candle is initially lit and it begins to burn, what prediction can you make and why?

Answers

1. Burning occurs when a chemical element or compound reacts with oxygen from the air through the aid of a flame.

2. A lack of oxygen and the depletion of all the wax are factors that would cause the burning process of a candle to stop.

3. It will continue to burn until there is a lack of oxygen or the wax is depleted because the reaction will continue until it cannot go further.

> The burning candle wax that is consumed in the chemical reaction with oxygen is converted into gaseous water, gaseous carbon dioxide, and similar products like carbonic acid.
>
> Now let's consider the burning of sugar. When table sugar burns in a pan it yields gaseous carbon dioxide and gaseous water. A certain amount of black carbon residue usually sticks to the bottom of the pan and interrupts the reaction, causing the burning process to end before all of the sugar is consumed.

Questions

1. What prediction can you make if black residue is collecting at the bottom of a pan of burning sugar?

2. What explanation can you give concerning why the black carbon residue stops the

burning process of sugar?

Answers

1. The table sugar will stop burning prematurely.

2. The black carbon insulates the remaining sugar and cuts it off from the heat source which is necessary to perpetuate the burning process.

We can always predict that the products that result from the burning reaction will contain oxygen. This is true because without the presence of oxygen, the burning reaction will not occur.

The burning of paper has many fascinating properties. Suppose we leave a sheet of paper on a table in an oxygen-filled room for many years. Suppose that we never apply a match to the paper to begin the burning process. However, when we return many years later, we find that the paper has turned yellow as if it had slowly burned. The paper has actually burned in some sense. Over time, the paper slowly reacts with oxygen in the surrounding air and undergoes a type of burning reaction without the presence of a flame. This demonstrates that, for some burning reactions, the flame only speeds up a burning reaction that would occur over time anyway.

Questions

1. What prediction can you make about paper that is left in an oxygen-rich environment with no flame applied to the paper?

2. What prediction can you make if a flame is applied to paper in an oxygen-rich environment?

Answers

1. The paper will turn yellow over a period of time in a gradual burning process.

2. The paper will burn rapidly.

READING A LINE GRAPH

A line graph is a picture that shows the relationship between two quantities. In a line graph, a line or several lines appear in a measured graph that coordinates a quantity on the horizontal direction (X axis) and a separate quantity on the vertical direction (Y axis). The line itself gives the actual simultaneous connection between the two quantities. The title expresses the information being conveyed by a line graph.

Familiarize yourself with the line graph shown on the following page.

The line graph depicted in Figure 1 contains information that demonstrates Charles' Law For Gases. Charles' Law For Gases states that there is a direct proportionality between temperature and volume (with pressure held constant). The fact that the lines for the different gases clearly have a constant slope (rise over run) demonstrates Charles' Law For

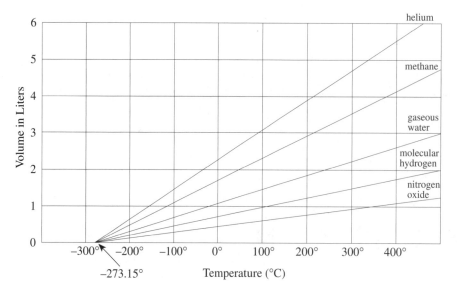

Figure 1: Charles' Law For Gases

Gases. For a gas depicted on the line graph, if we select a volume, then we can find the temperature. Conversely, if we select a temperature, then we can determine volume.

To find the volume of helium at 200 degrees Celsius, first you have to find 200 degrees Celsius on the X axis. Trace upwards from 200 degrees Celsius until you intersect the line that represents helium. Trace straight toward the left from the intersection point until you reach the Y axis. Read the volume that is listed at the point you have reached on the Y axis. In this case, the line graph indicates that helium has a volume of four liters.

Questions

1. What volume does gaseous water have when it is at a temperature of 300 degrees Celsius?

2. What temperature is methane when it is at a volume of four liters?

3. What gas has a volume of one liter at 400 degrees Celsius?

4. What is the obvious pattern or trend that is evident in this line graph?

Answers

1. 2 liters; to find the volume of gaseous water at 300 degrees Celsius, first you have to find 300 degrees Celsius on the X axis. Trace upwards from 300 degrees Celsius until you intersect the line that represents gaseous waters. Trace straight towards the left from the intersection point until you reach the Y axis. Read the volume that is listed at the point you have reached on the Y axis. In this case, the line graph indicates that gaseous water has a volume of two liters.

2. 400; to find the temperature of methane when it is at a volume of four liters, first you have to find four liters on the Y axis. Trace straight toward the right from four liters until you intersect with the line that represents methane. Trace down from the intersection point until you reach the X axis. In this case, the line graph indicates that methane has a temperature of 400 degrees Celsius when it is at a volume of four liters.

3. Nitrogen oxide; First, you have to find 400 degrees Celsius on the X axis. Then, you have to find one liter on the Y axis. Trace upwards from 400 degrees Celsius and trace straight toward the right from one liter until you reach an intersection point. Read the gas that is listed at the intersection point you have reached. In this case, the line graph indicates that nitrogen oxide has a volume of one liter at 400 degrees Celsius.

4. Increasing temperatures of gas result in an increase of the volume of the gas.

TOPICS IN CHEMISTRY

This material will give you an overview of the chemistry field. These are topics you will need to become familiar with before taking the GED examination.

Matter

Matter is everything that has volume and mass. Water is matter because it takes up space, light is not matter because it does not take up space.

States of Matter

Matter exists in three states, as follows:

State	Properties	Example
solid	definite volume, definite shape	ice
liquid	definite volume, no definite shape	water
gas	no definite volume, no definite shape	water vapor or steam

Thermal energy causes molecules or atoms to vibrate. As vibration of particles increases, a material may change to a different state; it may melt or boil. Decreasing energy in a material may cause condensation or freezing. Another name for thermal energy is heat. Temperature is a measure of the average kinetic energy, or vibration, of the particles of a material. For most materials, the boiling point and freezing point are important. The boiling point of water is 100°C, and its freezing point is 0°C.

Structure of Matter

Atoms are the basic building blocks of matter. Atoms are made of three types of subatomic particles, which have mass and charge. Protons and neutrons are found in the nucleus, or solid center of an atom. Electrons are found in the outer portion of an atom. This outer portion is mostly made of empty space. Under most conditions, atoms are indivisible. Atoms may be split or combined to form new atoms during atomic reactions. Atomic reactions occur deep inside the sun, in nuclear power reactors and nuclear bombs, and in radioactive decay.

Subatomic Particle	Mass	Charge	Location
proton	1 amu	+1	nucleus
neutron	1 amu	0	nucleus
electron	0 amu	−1	outside nucleus

Most atoms have equal numbers of protons and electrons, and therefore no net charge. Atoms with unequal numbers of protons and electrons have net positive or negative charges. Charged atoms are called ions. Atomic mass is determined by the number of protons and neutrons in an atom. The way to express atomic mass is in atomic mass units (amu).

State Change	Process Name	Heat Change
solid ⟶ liquid	melting	heat added
liquid ⟶ gas	evaporation or boiling	heat added
gas ⟶ liquid	condensation	heat removed
liquid ⟶ solid	freezing	heat removed

A material made of just one type of atom is called an element. Atoms of an element are represented by symbols of one or two letters, such as C or Na. Two or more atoms may combine to form molecules.

Atoms of the same element have the same number of protons in their nucleus. An atom is the smallest particle of an element that retains the characteristics of that element. Each element is assigned an atomic number, which is equal to the number of protons in an atom of that element. The Periodic Table is a chart listing all the elements in order according to their atomic number. The elements are grouped vertically in the Periodic Table according to their chemical properties. The Periodic Table is a reference tool used to summarize the atomic structure, mass, and reactive tendencies of elements.

Molecules are clusters of atoms. Molecules form, decompose, or recombine during chemical reactions. Materials made of one type of molecule are called compounds. Compounds may be represented by formulae using atomic symbols and numbers. The numbers show how many atoms of each type are in the molecules. For example, the symbol for water, H_2O, shows that a molecule of water contains two hydrogen atoms and one oxygen atom. Atomic symbols without subscript numbers represent just one atom in a molecule.

Chemical compounds containing carbon are called organic, because these materials are often made by living things. The chemistry of organic compounds is complex and distinct from that of other compounds. Therefore, organic chemistry is a large and distinct discipline. Compounds without carbon are called inorganic.

Mixtures are materials made of two or more compounds or elements. They can be separated by physical means, such as sifting or evaporation. Liquid or gas mixtures are called suspensions, colloids, or solutions. Suspensions have particles that settle out unless the mixture is stirred. Dust in air is a suspension. Colloids have particles large enough to scatter light, but small enough to remain suspended without stirring. Milk is a colloid; it is opaque because its particles scatter light. Solutions have particles so small they do not scatter light. They are transparent to light and their particles do not settle out.

Chemicals that are dissolved in solutions are called solutes. A substance that dissolves another to form a solution is a solvent. In a salt water solution, water is the solvent. Not all chemicals can function as solvents. Some solvents (like gasoline) are able to dissolve only certain solids. Water is sometimes called the "universal solvent" because it is able to dissolve so many chemicals.

Concentration is a measure of how much solute is in a solution. A given amount of solvent is able to dissolve only a limited amount of solute. This amount may be increased if the solution is heated or pressure on the solution is increased. Dilute solutions have relatively little solute in solution. Concentrated solutions have a lot of solute in solution.

Solutions that are able to dissolve more solute are called unsaturated. Solutions that cannot dissolve more solute are called saturated. Solutions that are saturated at high temperature or high pressure may become super-saturated at lower temperatures or pressures. Supersaturated solutions contain more dissolved solute than normally is present in a saturated solution. These solutions are unstable, and solute may crystallize out of the solution easily.

Chemical Reactions

Matter may undergo chemical and physical changes. A physical change affects the size, form, or appearance of a material. These changes can include melting, bending, or cracking. Physical changes do not alter the molecular structure of a material. Chemical changes do alter the molecular structure of matter. Examples of chemical changes are burning, rusting, and digestion.

Under the right conditions, compounds may break apart, combine, or recombine to form new compounds. This process is called a chemical reaction. Chemical reactions are described by chemical equations, such as NaOH + HCl $\longrightarrow$ NaCl + H_2O. In a chemical equation, materials to the left of the arrow are called reactants and materials to the right of the arrow are called products. In a balanced chemical equation, the number of each type of atom is the same on both sides of the arrow.

unbalanced: $H_2 + O_2 \longrightarrow H_2O$

balanced: $2H_2 + O_2 \longrightarrow 2H_2O$

There are four basic types of chemical reactions: synthesis, decomposition, single replacement, and double replacement. A synthesis reaction is one in which two or more chemicals combine to form a new chemical.

EXAMPLE

$A + B \longrightarrow AB$, or $2H_2 + O_2 \longrightarrow 2H_2O$

A decomposition reaction is one in which one chemical breaks down to release two or more chemicals.

EXAMPLE

$AB \longrightarrow A + B$, or $2H_2O \longrightarrow 2H_2 + O_2$

A single replacement reaction involves a compound decomposing and one of its constituent chemicals joining another chemical to make a new compound.

EXAMPLE

$AB + C \longrightarrow A + BC$, or $Fe + CuCl_2 \longrightarrow FeCl_2 + Cu$

A double replacement reaction is one in which two compounds decompose and their constituents recombine to form two new compounds.

EXAMPLE

$AB + CD \longrightarrow AC + BD$, or $NaOH + HCl \longrightarrow NaCl + H_2O$

Acids and Bases

Acid and base are terms used to describe solutions of differing pH. The concentration of a hydrogen ion in a solution determines its pH, which is based on a logarithmic scale.

Solutions having pH 0–7 are called acids and have hydrogen ions (H+) present. Common acids include lemon juice, vinegar, and battery acid. Acids are corrosive, and taste sour. Solutions pH 7–14 are called bases (or alkaline), and have hydroxide ions (OH−) present. Bases are caustic and feel slippery in solution. Common bases include baking soda and lye. Solutions of pH 7 are called neutral and have both ions present in equal but small amounts.

The reaction created when an acid and base combine is a double replacement reaction known as a neutralization reaction. In a neutralization reaction, acid + base $\longrightarrow$ water + salt. Figure 2 is an example.

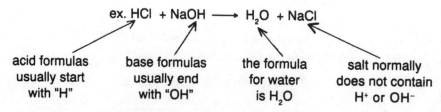

Figure 2: Example of a Neutralization Reaction

PHYSICAL SCIENCE

Measurement

The physical characteristics of an object are determined by measurements. Measured characteristics include mass, volume, length, temperature, time, and area. There are two common measurement systems, English and metric.

The English system, used most often in the United States, does not have a consistent system of conversion factors between units.

EXAMPLE

1 yard = 3 feet, 1 foot = 12 inches,
1 yard = 36 inches

The metric system, used most often in science, has conversion factors between units based on multiples of 10.

EXAMPLE

1 kilometer = 1,000 meters,

1 meter = 100 centimeters
= 1,000 millimeters

Prefixes in the metric system indicate the number of multiples of the base units, so it is simple to determine the conversion factors between units.

A third measurement system, the International System of Units, is based on the metric system. The International System of Units differs from the metric system by using the Kelvin temperature scale. The size of a degree on the Celsius and Kelvin scales is the same, but "0°" is different. 0 Kelvin = –273°Celsius. 0 Kelvin, also known as absolute zero, is the temperature at which, theoretically, all molecular movement ceases.

Characteristic	English System	Metric System
mass/weight	pound (weight)	kilogram (mass)
volume	quart	liter
length	foot	meter
temperature	°Fahrenheit	°Celsius
time	second	second

Prefix	Multiplication Factor	Unit Symbols
kilo	× 1000	km, kg, kl
no prefix (base unit)	× 1	m, g, l
deci	× 0.1	dm, dg, dl
centi	× 0.01	cm, cg, cl
milli	× 0.001	mm, mg, ml

To convert from °Celsius to °Fahrenheit, use the following equation:

$$°C = \frac{5}{9}(°F - 32).$$

To convert from °Celsius to Kelvin, use the following equation:

$$°C + 273 = K.$$

☞ Practice: Chemistry

> **DIRECTIONS:** Read the following passages. Apply the strategies you've learned. Write in the margins and mark up the text as you go. Then answer the questions after each passage.

Questions 1–5 refer to the following passage.

The beginning of the twentieth century saw the development of a new kind of chemistry called **nuclear chemistry**. **Nuclear chemistry** is the study of the parts of, and changes within, the heart of the atom called the **nucleus**.

Uranium, the most common fuel used in nuclear fusion reactors, was discovered in 1789. However, it was not known to be radioactive and dangerous. People actually carried pieces of it in their pockets and developed radiation burns from it! The first discovery of particles being emitted from uranium was made by Henri Becquerel in 1895. This discovery was followed shortly by the discovery of other radioactive elements—polonium and radium.

Nuclear fission is the splitting apart of the nucleus of an atom. Tremendous amounts of radiation are released when the nucleus is split. This process was accidentally discovered by Otto Hahn and Lise Meitner in 1932. Ten years later, Italian physicist Enrico Fermi successfully built the first modern, controlled device for producing nuclear fission. The device, later called a **reactor**, was called a pile because it was completely surrounded by a pile of lead bricks. The lead bricks were necessary to contain the radiation that was released as a result of nuclear fission.

Reactors and similar research equipment permit us to study nuclear chemistry in a clear and detailed fashion. Today, large fission reactors produce electricity. In France, over half of the electricity that is used is produced from nuclear fission. There is a substantial amount of radioactive material that results from nuclear power production.

The biggest single science project in the history of modern man was the Manhattan Project, a secret U.S. Government program to develop the atomic bomb during World War II. Fission atomic bombs work on the chain-reaction principle. As an atom breaks apart, neutrons are released, creating consecutive nuclear reactions, so that the bomb's impact grows greater. Explosive nuclear fission was used to develop an even more powerful hydrogen bomb. The hydrogen bomb uses the fission bomb's chain-reaction as a trigger to join hydrogen atoms (fusion) that results in a large explosion. Each of the original hydrogen bombs had the power of about 10,000 tons of TNT.

The development of the fission and hydrogen bombs has had a profound impact on humanity. For the first time in human history, man was able to destroy virtually all life on Earth.

1. Circle the sentence that best represents the main idea of the preceding passage.

    (A) Nuclear chemistry is a dangerous field of study that involves nuclear waste and radiation.

    (B) The study of nuclear chemistry led to the development of nuclear power and nuclear bombs.

    (C) Nuclear chemistry is more interesting than basic chemistry.

    Fill in the blanks.

2. _____ is the most commonly used nuclear fuel.

3. A device that was originally called a pile is now known as a _____.

4. Nuclear chemistry is the study of the parts of and the changes in the heart of an atom, called the _____.

5. The Manhattan Project was undertaken by the U.S. government to develop _____.

Questions 6–7 refer to the following passage.

   All reactions need to receive a certain amount of energy before they can start. The amount of energy needed or received to start the chemical reaction is called activation energy. Some reactions require so little energy that it can be absorbed from the surroundings. This is called a spontaneous reaction, which takes place with so little energy that it seems as if no energy was needed. A reaction that gives off energy is called an exothermic reaction. A reaction that absorbs energy is called an endothermic reaction. Combustion is a decomposition reaction. A catalyst can be added to a chemical reaction to speed up or slow down the reaction rate.

6. What is the difference between an endothermic reaction and an exothermic reaction?

   _____
   _____
   _____

7. What do a spontaneous reaction, an exothermic reaction, and an endothermic reaction all have in common?

   _____
   _____
   _____

Fill in the blanks.

8. A suffix is placed _____ the main chemical name.

9. A prefix is placed _____ the main chemical name.

10. The prefixes of alkanes that are higher than four use traditional Latin _____.

Circle "T" if the statement is true; circle "F" if the statement is false.

11. Methane has nine carbon atoms.

    T   F

12. Prefixes tell you how many carbon atoms are in an alkane.

    T   F

13. All alkanes have the same suffix.

    T   F

14. Octane has eight carbon atoms.

    T   F

GED Science

Break the following alkane down into a prefix and a suffix.

15. Pentane: _____ _____

16. Pent- is the Latin root for the number _____.

Questions 17–19 refer to the following passage.

Corrosive rusting of iron and steel is another process involving the combination of a material with oxygen that results in the material being transformed into different chemical compounds. It is similar to burning, but requires moisture to be present in the oxygenated air. In an extremely dry environment, rusting occurs very slowly or does not occur at all. In a moisture-laden environment that has little oxygen, rusting will also occur very slowly or not at all. If both oxygen and moisture are present in sufficient amounts, then a chain of reactions will occur. Iron is dissolved and becomes charged and goes through a chemical reaction that results in rust and water. The resulting water helps to continue the rusting reaction until all of the iron has been transformed into rust.

17. What prediction can you make if a piece of iron is left in a totally dry environment?

_____
_____
_____

18. What prediction can you make if a piece of iron is left in an environment that has no oxygen?

_____
_____
_____

19. What prediction can you make if a piece of iron is left in an environment that is moisture-laden and rich with oxygen?

_____
_____
_____

Answers

1. (B)

2. Uranium

3. reactor

4. nucleus

5. the atomic bomb

6. An endothermic reaction *absorbs* energy while an exothermic reaction *gives off* energy.

7. They all need a certain amount of energy before the reactions can start.

8. at the end of

9. at the beginning of

10. roots

11. False. Methane has one carbon atom. Nonane has nine carbon atoms.

12. True. Alkanes have prefixes in their chemical names that indicate the number of carbon atoms appearing in the compound.

13. True. All alkanes have "ane" as their suffix.

14. True. The prefix "oct" indicates there are eight carbon atoms in octane.

15. "Pent" is the prefix; "ane" is the suffix.

16. 5

17. The iron will rust very slowly or won't rust at all.

18. The iron will not rust at all because oxygen is needed for rusting to occur.

19. The iron will rust relatively quickly and eventually be turned entirely into rust.

REVIEW

In this chapter we learned how to make predictions, find main ideas, and compare and contrast information. We also learned how to understand line graphs, chemical formulas, and prefixes and suffixes.

Understanding line graphs and chemical formulas, as well as learning prefixes and suffixes, will help you develop a basis for the skills that will be needed when attacking chemistry questions on the GED exam. These are extremely useful tools that are especially vital for chemistry problems.

Finding main ideas is important because it helps us get the full picture by sorting through and arranging all the details of a passage.

When we compare and contrast, we see relationships. Understanding the relationship between two things will help us understand the important facts of scientific material. If we understand the relationship between two items, then we will have a good understanding of the two items themselves. By understanding relationships, we can then learn how to make predictions since we see how one thing affects another.

Making predictions helps us apply our knowledge and serves as a good test of ourselves to see how well we understand what we have learned. We draw all our skills together when we have to make a prediction.

Science

Earth Science

SCIENCE

EARTH SCIENCE

READING A MAP

Being able to read a map correctly is a skill that can be useful in many situations that life presents. You might find yourself in an unfamiliar area with no one available to give you directions. You might want to check the weather report in the newspaper. You might need to read a map in order to fulfill your job duties. These are only a few examples of the many situations in life that require map reading skills.

Reading a map can seem like an overwhelming task when you first look at one. There are many different kinds of maps detailing a wide variety of information. The easiest way to overcome the feeling that reading a map is overwhelming is to look at different parts of the map as individual pieces of information. Once you have broken the map down into individual pieces, the information becomes easier to comprehend. Once you have comprehended the individual pieces of information that make up a map, you can then put them together and read the whole map.

Although there are many different kinds of maps with a wide variety of information, most maps have common characteristics. The first piece of information that is important to look at on a map is the title of the map. The title will tell you the type of information that you can expect to obtain from the map. For example, a map detailing the weather forecast for the following day might be titled "Tomorrow's Weather Forecast." The title of a map is usually located at the top of the map.

The second piece of information that is important to look at on a map is the map key. The map key will tell you specific information about the contents of the map. The map key is usually located on the edge of a map, often in a corner. The map key can either be labeled as a "map key" or it can be a group of symbols

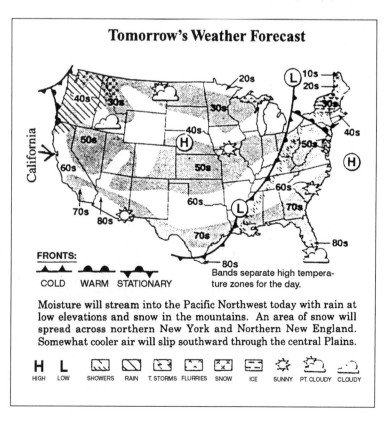

with accompanying words that explain what the symbols represent. Take a look at the map shown on the previous page.

Locate and circle the title. Locate and circle the map key. The title clearly states that the map has information about tomorrow's weather forecast. The map key is located below the map. The map key on this map is broken into two parts. The two parts of the map key are separated by a paragraph that explains, in writing, the information that is shown on the map. The top part of the key is labeled "fronts." Notice that the three different types of fronts are distinguished by differently shaped symbols attached to a line.

Locate and circle a warm front on the map. Locate and circle a cold front on the map.

Questions

1. What symbol is associated with a warm front? _____

2. What symbol is associated with a cold front? _____

Answers

1. A warm front is distinguished by a half circle by the map key.

2. A cold front is distinguished by a triangle by the map key.

The bottom part of the key does not have a label. However, there is a descriptive word below each symbol that clearly states the meaning of the symbol. Find the symbol that represents sunny weather. Now locate the same sunny weather symbol on the map. One quick glance and you now know that, in the area around the "sunny" symbol, tomorrow's forecast is for sunny weather.

Find the symbol that indicates rainy weather. Find the symbol that represents showery weather. What is the difference between these two symbols? Notice that the rainy symbol is represented by unbroken slanted lines, while the showery symbol is represented by similar slanted lines that are broken in the middle. Locate the rainy symbol on the weather map. Locate the showery symbol on the weather map. Once again, by interpreting the rainy and showery symbols you can quickly determine that the area covered by the symbol can expect rainy or showery weather.

There are two symbols at the left end of the lower map key: "H" for high and "L" for low. These symbols represent weather phenomena called high pressure and low pressure. High pressure generally indicates clear, sunny weather while low pressure generally indicates rainy or snowy weather. Fronts also generally indicate rainy or snowy weather. However, you might find the high and low pressure and the information about the fronts irrelevant since the other symbols clearly indicate where sunny weather, cloudy weather, rainy weather, and other weather phenomena are forecast to occur tomorrow.

Earth Science

What other information can be obtained from the map forecasting tomorrow's weather that is not contained within the map key? Notice that there are shaded and unshaded bands that have numbers within them; for example, 40s, 50s, and 60s. Find the shaded band that has the number "50s" within it. Notice that the band stretches all the way across the country. All of the areas of the country that are in this shaded band can expect temperatures in the 50s tomorrow.

As you can see from this lesson, the weather map indicating tomorrow's weather forecast might look overwhelming at first. But by locating the map key, reading the symbols individually, and locating the symbols on the map, you can easily determine the weather expected in any part of the country.

Questions

Refer to Figure 1 to answer the following questions.

1. Is any rain forecast to fall in any part of California? _____

2. What daytime temperatures are forecast throughout the entire state of California? _____

3. Now find the state that you live in. See if you can determine what type of weather and daytime temperatures are forecast for your state. Write your answers below:

   Weather: _____

   Temperature: _____

Answers

1. Yes; there are lines indicating there will be showers.

2. The daytime temperatures forecast throughout the entire state of California are 50s, 60s, 70s, and 80s.

3. First, find the state you live in. The next step is to determine what type of weather and daytime temperatures are forecast in your state by using what you have learned in this section. For example, pretend you live in New Jersey. The forecast for the weather in New Jersey is clear, and the forecast for daytime temperatures is in the 50s.

Figure 2 represents a local weather map covering the state of New Jersey and portions of the states of New York and Pennsylvania. Notice that the map is broken into three distinct regions that are located on both sides of the warm and cold fronts. The freezing point of water is 32 degrees Fahrenheit. Below 32 degrees Fahrenheit, you can expect frozen precipitation to be occurring, usually as snow. Above 32 degrees Fahrenheit, you can usually expect rain to be occurring.

Questions

The triangles on the cold front and the half-circles on the warm front identify the type of front that is depicted. These symbols also face in the direction that the front is moving.

1. Toward what direction is the cold front moving? _____

2. Toward what direction is the warm front moving? _____

3. What is the temperature range in Region B? _____

4. What is the temperature range in Region C? _____

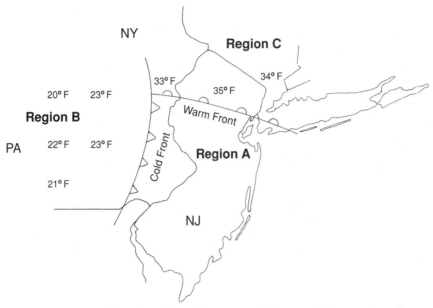

Figure 2: A Local Weather Map for New Jersey and Parts of Pennsylvania and New York

5. What type of precipitation would likely be occurring in Region B? _____

6. If you had to drive in Region C, would you expect the roads to be icy? _____

Answers

1. East; the triangles on the cold front identify the type of front that is depicted. The triangles are facing east. This means the cold front is moving east.

2. North; the half-circles on the warm front identify the type of front that is being depicted. The half circles face in the direction the front is moving. On this map, the half-circles are facing north. This means the warm front is moving north.

3. 20–23° F

4. 33–35° F

5. Snow; the temperatures in Region B are below 32 degrees Fahrenheit. Below that temperature, you can expect frozen precipitation to occur, usually as snow.

6. No; the temperatures in Region C are above 32 degrees Fahrenheit, the freezing point.

Take a look at the map on the following page (Figure 3) and read the meanings of the symbols represented on it.

Questions

1. What is the main source of diamonds?
   _____
   _____
   _____

2. What would be a good title for this map?
   _____
   _____
   _____

3. Does Australia have any oil deposits?

Earth Science

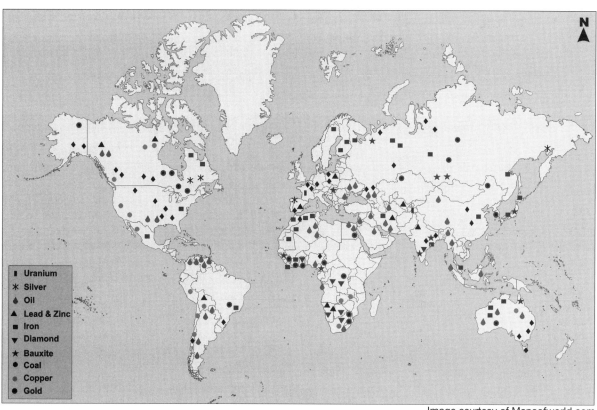

Figure 3

Image courtesy of Mapsofworld.com

4. What is the one mineral shown to be in supply in Scandinavia?

Answers

1. The African continent.

2. A good title would be, "World Map of Minerals." The symbols on the map represent the locations of various minerals around the globe.

3. Yes, oil is signified by the "drop" symbol in three locations.

4. Iron, specifically in Finland and Sweden.

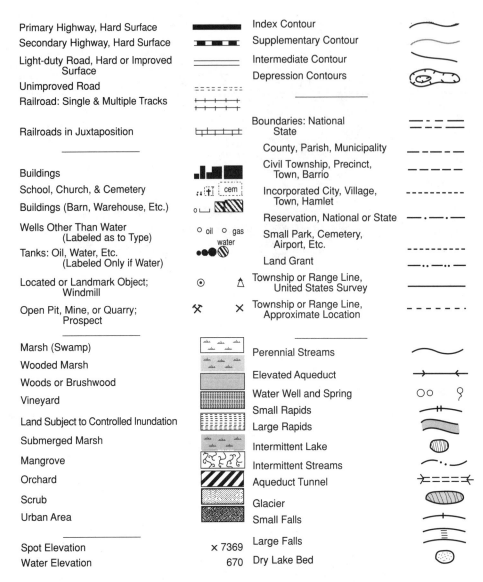

Figure 4: A Topographic Map Key

Topographic maps are extremely useful in the study of earth science. The map key that is shown above explains the information that can be obtained from a topographic map.

Question

1. If you were planning a class trip that would discuss subjects involving earth science topics, how much of the information from the topographic map key could you use to plan your trip?

Answer

1. Highway information; pit, mine, and quarry information; contours, park boundaries, stream and lake information.

In addition to providing information about natural resources and human constructions, topographic maps also provide information about the elevation of land. Take a look at the symbols on the top right portion of the topographic map key. These symbols represent contours on a topographic map. A **contour** is a line that represents a common elevation in feet at numerous locations on a topographic map. An **index contour** represents common elevations at 100-feet intervals. An **intermediate contour** represents common elevations at 20-feet intervals.

Contours that are close together indicate a sharp change in elevation. The contour with the hash marks represents a place on the map where a depression (reduction) in elevation is located. Take a look at the symbols on the bottom left portion of the topographic map key. These symbols represent the elevation in feet at a place on a topographic map. Such spot elevations are often given for mountain peaks. Refer to the topographic map on the following page (Figure 5).

Questions

1. What is the maximum elevation of Sargent Hill? _____

2. What do the tight contour lines on Scott Mountain represent as compared to Sargent Hill? _____

3. On a map such as this topographic map, which direction is which, and how can one tell? _____

4. Predict why the Mettawee Valley Cemetery is located where it is on this map. _____

5. Based upon information provided by the contours, what characteristic does Town Hill have that Sargent Hill does not share? _____

Answers

1. 1,087 feet

2. The closeness of the contour lines shows Scott Mountain to be a steep terrain as compared to Sargent Hill. The close proximity of the lines indicates a rapid change in elevation on the mountain. The wider spaces between the lines on the hill indicate a more gradual change in elevation.

3. Unless otherwise indicated, the top of a map will represent north. West is left, east is right, and, of course, south is at the bottom of the map. A majority of maps of all types will follow this general rule.

4. Due to the lack of contour lines in this area of the map, it can be assumed that the topography is level and unobstructed. There is also no river or stream in the direct area so flooding from the hills is not a concern.

5. Town Hill has multiple peaks.

Take a moment to familiarize yourself with the map shown on page 87 (Figure 6). Locate the key and read the names of the ocean currents listed on the map.

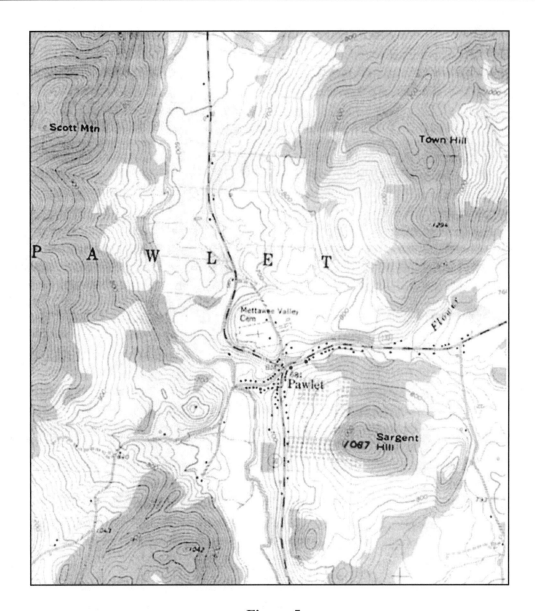

Figure 5

Questions

1. What would be a good title for this map? _____

2. The Gulf Stream current is warm. Toward which continent does the Gulf Stream current flow? _____ What effect would you expect this current to have on the continent named above? _____

3. If you released a bottle into the North Pacific Drift current from the coast of California, where would the bottle end up?

Earth Science

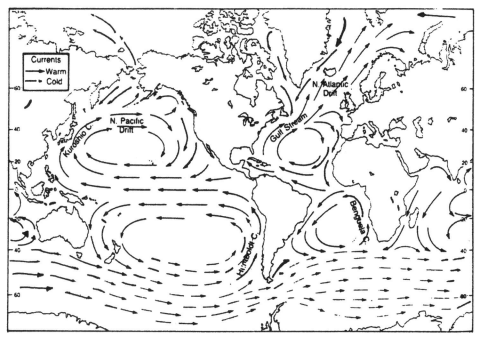

Figure 6

Take a moment to familiarize yourself with the map shown on the following page (Figure 7). Locate the title of the map and the map key.

Answers

1. Global Ocean Currents; the map represents the world and its ocean currents.

2. Europe; it warms the waters along the ocean floor.

3. Asia, then, possibly, back on the West Coast of the United States.

Questions

1. If you wanted to open a business that needed a local supply of anthracite coal, in which state would you open your business?

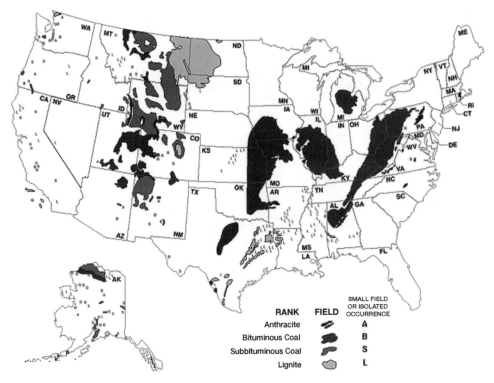

Figure 7

2. If the environmental authorities in your state required you to burn lignite coal in your power plants because it was a cleaner-burning coal, from which states would you obtain this type of coal?

3. Which type of coal appears to be present across the largest area of land?

Answers

1. Pennsylvania; anthracite, represented mostly in black, is found mostly in Pennsylvania.

2. North and South Dakota, Montana, Wyoming, Utah, Colorado, New Mexico, Arizona.

3. Bituminous; there is more bituminous represented on this map than anthracite or lignite.

Take a moment to familiarize yourself with the map shown on the following page (Figure 8). Find the map key and note the risks of damage from an earthquake and how the risks are represented.

Earthquakes occur when breaks in the Earth's crust, called **faults**, move and release energy. The energy that is released in an earthquake is caused by friction that builds up when adjoining sections of the Earth's crust, called **plates**, rub along one another. An example of this stored energy is the friction between two blocks of rough wood. Put them together and

Earth Science

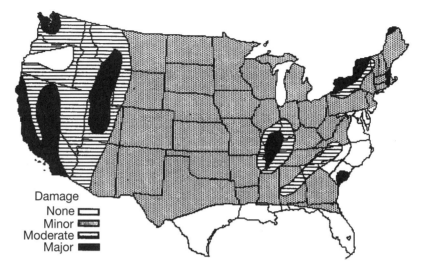

Seismic-risk map for the U.S. showing earthquake damage ares of reasonable expectancy in the next 100 years (Derived from U.S. Coast and Geodetic Survey, ESSA Rel. Es-1, January 14, 1969

Figure 8

try to make one slide over the other. It takes a hard push before friction is overcome. One block then moves a short distance before it "catches" again and won't move farther until another hard push is applied. Like one rough block trying to slide along the other, the plates jarringly move past each other and release a tremendous amount of energy that causes the earth to shake. The amount of shaking that occurs depends directly on the amount of friction that is overcome when the plates move.

There are many local geologic factors that tend to increase or diminish the strength and subsequent damage caused by an earthquake. The depth at which an earthquake occurs below the surface also directly influences the amount of damage it causes. As a general rule, a major damaging earthquake occurs when a large strain and subsequent release of energy originates near the Earth's surface.

Questions

1. What state appears to have the greatest risk of having earthquake damage during the next 100 years? _____

2. Name two other states that have a great risk of having earthquake damage during the next 100 years. _____

3. What is the expected risk of damage from earthquakes over the majority of the United States during the next 100 years? _____

4. Name two states in which there is no risk of damage from earthquakes during the next 100 years. _____

5. Which state has both a major risk and no risk of damage from an earthquake during the next 100 years? _____

Answers

1. California; major damage is represented on the map in black. Most of California is in black.

2. New York, Washington; a large area of Washington and New York is in black.

3. Minor; although large sections of California, New York, and Washington are in black, representing major damage, the majority of the map represents minor damage.

4. Delaware, Maryland; Delaware and Maryland are completely in white, which represents no damage.

5. South Carolina; one section of South Carolina is in black, representing major damage; the other section is in white, representing no damage.

Take a moment to familiarize yourself with the map of the continents of North and South America shown below (Figure 9).

The numbers along the left edge of the map represent latitude. The numbers along the top edge of the map represent longitude. Latitude and longitude are used to determine locations, called coordinates, on the Earth's surface. Latitude lines run horizontally parallel to the equator. Longitude lines run vertically and run through the Earth's north and south poles.

Figure 9: North and South America

The unit of measurement for latitude and longitude lines is degrees, with 360 equal degrees in a full circle. The latitude and longitude lines on this map are drawn at 20-degree intervals. A single degree is sometimes divided into 60 equal parts called minutes, and minutes can be broken down into 60 equal parts called seconds. However, on this map only degrees of longitude and latitude are represented. Notice that the latitude lines above the equator are marked "N" for north, while the latitude lines below the equator are marked "S" for south. The longitude lines to the left of the Greenwich Line, which runs through Greenwich, England are marked "W" for west. Those marked "E" for east are located to the right of the Greenwich Line. The International Date Line passes over the middle of the Pacific Ocean and can be labeled either 180 degrees west longitude or 180 degrees east longitude.

Questions

1. What country is located at 100 degrees west longitude and 60 degrees north latitude? _____

2. What country is located at 100 degrees west longitude and 40 degrees north latitude? _____

3. What country is located at 60 degrees west longitude and 0 degrees (equator) latitude? _____

4. What country is located at 60 degrees west longitude and 40 degrees south latitude? _____

5. If you were told to start at 80 degrees west longitude and 0 degrees (equator) latitude, and to fly to 40 degrees longitude, what country would you end up in? _____

6. Would you be able to precisely track a

ship sailing from Argentina to the United States by using the intervals of longitude and latitude given on this map? Explain.

_____

_____

_____

Answers

1. Canada; find 100 degrees west longitude on the top edge of the map, and find 60 degrees north latitude. You can pinpoint Canada at this location.

2. United States; find 100 degrees west longitude on the top edge of the map, and find 40 degrees north latitude. You can pinpoint the United States at this location.

3. Brazil; find 60 degrees longitude west on the top edge of the map, and find 0 degrees (equator) latitude. You can pinpoint Brazil at this location.

4. Argentina; find 60 degrees west longitude on the top edge of the map, and find 40 degrees south latitude. You can pinpoint Argentina at this location.

5. United States; find 80 degrees west longitude on the top edge of the map, and find 0 degrees (equator) latitude. Then from this location, trace your finger on the map to 40 degrees north latitude. You can pinpoint the United States at this location.

6. No. Since the ship would not be sailing directly along the lines on this map we could not precisely track it. One would need a map with all the lines of longitude and latitude in order to pinpoint the ship's position during every part of its journey.

Take a moment to familiarize yourself with the map shown below (Figure 10). Notice that there is no key for this map. The title of the map contains enough information to explain the contents of the map (As used on this map, "mean" signifies "average.")

Questions

1. Which mean sea-level temperature line passes through central Florida? _____

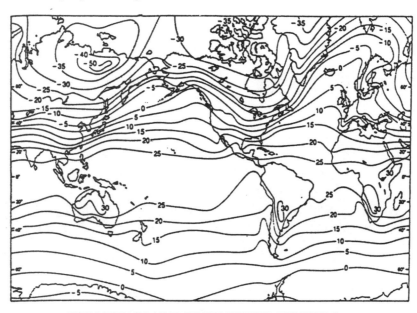

WORLD MEAN SEA-LEVEL TEMPERATURES IN JANUARY IN °C

Figure 10

2. Follow the mean sea-level temperature line that goes through Florida. What other continents does this line cross? _____

3. Compare the mean sea-level temperature lines in the northern part of the Atlantic Ocean along the West Coast of Europe with the previous ocean current map. Do you notice a correlation between the mean sea-level temperatures and the Gulf Stream current? Explain.

    _____
    _____
    _____

Answers

1. 15°C; find central Florida on the map. If you trace the line going through central Florida in either direction, you will see it is 15 degrees Celsius.

2. Africa, Asia; if you continue to trace your finger along the line, you will see it goes through Africa and Asia.

3. Yes; mean sea-level temperatures appear to be higher along the West Coast of Europe, the same place as the Gulf Stream current.

UNDERSTANDING COMPOUND WORDS

Understanding compound words is a critical skill that can be used effectively to understand scientific words. A compound word is simply the combination of two words into a larger word. If you can understand the smaller words, then you can understand the larger compound word.

Snowstorm is an example of a scientific compound word. If you know what snow is and you know what a storm is then you can combine the meanings of the words. Knowing the meanings of these separate words, you can conclude that snowstorm is a storm in which snowfall occurs. List another scientific compound word: _____

Scientific words are often compound words. Therefore, the ability to understand scientific words is often as simple as breaking the word down into the two separate words, interpreting the separate words' meanings, and combining the two meanings into a definition that makes sense.

Complete the scientific compound word exercises that follow. Be sure to do all of the exercises even if you find the scientific compound words to be words that you already understand. These exercises will help you get in the habit of remembering to break unknown scientific compound words into smaller words.

Questions

Scientific compound word exercise.

1. Earthquake _____ _____

    Meaning: _____
    _____

2. Waterfall _____ _____

    Meaning: _____
    _____

3. Sandstone _____ _____

    Meaning: _____
    _____

4. Thunderstorm _____ _____

    Meaning: _____
    _____

Answers

1. earth quake

    Planet humans inhabit, shaking

2. water fall

    water in a falling motion

3. sand stone

    sand in a stone form

4. thunder storm

    thunder associated with a rain storm

As you can see from the above examples, scientific compound words can come in a variety of forms. A natural phenomenon like an earthquake, a water formation like a waterfall, a rock like a sandstone, and a weather phenomenon like a thunderstorm can all be broken down into more understandable separate words.

FINDING DETAILS

Details can hold the key to understanding a scientific concept, theory, or presentation. For example, just looking at a weather map to see if it will be a sunny day does not give you enough information to determine if it would be a good day to go to the beach. Locating and interpreting the details on the weather map is crucial for making a decision about whether or not it is a good day to go to the beach. It might be a sunny day, but a closer examination of the map might reveal that the high temperature will only be in the low 60s and there will be a 30 mile-per-hour wind at the beach. Most people would decide not to make a trip to the beach on such a day, based upon the more detailed information.

The way to find details in a scientific concept, theory, or presentation is to read and/or examine it carefully. As you carefully read or examine it, make mental notes about the meaning of each piece of the concept, theory, or presentation. If some part of a concept, theory, or presentation does not make sense to you, isolate it to try to make sense of the part aside from the main idea. Remember, finding details can be crucial to understanding the concept, theory, or presentation as a whole.

Solar panels are used to make the concept of capturing energy from sunlight to provide heat a reality. The details of how a solar panel turns sunlight into heat are shown in the diagram shown on the following page (Figure 11).

These diagrams show how two different types of solar heating systems work. By looking at the diagram, we see that there is a lot more to turning sunlight into usable heat than just the solar panels. The information presented might seem overwhelming at first. Remember, read the information closely and try to interpret the different pieces of information separately before trying to understand the entire presentation.

The first piece of information that you would want to examine is the diagram of the solar collector (panel) on the roof, labeled "Solar Collector on Roof of House." You would start with this piece of information because it is probably the piece that you are most familiar with and is ultimately the source of the usable heat. By reading the information around the diagram you can determine that the diagram is a cross-section that shows pipes containing circulating water inside the solar collector. You can also see that the solar collector is attached to the roof and that there is a glass plate to keep heat from escaping on windy days. The diagram beneath it shows the solar collectors on the roof and hot water storage in the basement, but does not contain any details about how the system works.

The two diagrams, labeled "Solar Air Sys-

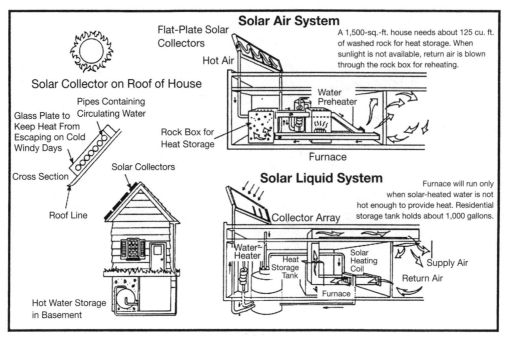

Figure 11

tem" and "Solar Liquid System," contain details about a solar system used to provide heat for a house or a building. Refer to these diagrams when answering the following questions.

Questions

1. What kind of heat storage device does the solar air system have?

2. What storage capacity is used in the solar liquid system?

3. What do both systems have for backup heating needs in case there is not enough sunlight to provide adequate heat?

4. How can you tell the flow direction of air or water in the systems?

Answers

1. Rock box; if you examine the diagram for the solar air system, you will notice an arrow pointing to a rock box for heat storage.

2. 1,000 gallons; it states in this section that, *Residential storage tanks hold about 1,000 gallons*.

3. Furnace; both the solar air system and the solar liquid system have a furnace for backup heat.

4. Arrows; for example, there are arrows labeled "return air" and "supply air."

Carefully read the passage below and answer the questions that follow. Remember to pay attention to details.

Earthquakes can cause devastating damage and loss of life. The strength of earthquake shock waves, which spread out from the earthquake's point of origin, called the epicenter, is recorded by seismographs. Seismologists, scientists who study earthquakes, interpret the recordings made on seismographs to determine the strength of an earthquake. As a general rule, the strength of an earthquake and subsequent damage decreases with distance away from the epicenter. However, a strong earthquake can cause damage over a wide area.

Seismologists use a number system called the Richter scale to rank the strength of an earthquake. It starts with zero and has no theoretical end. Each increase of one whole number on the Richter scale equals the release of 31 times more energy from an earthquake.

Earthquakes that have a Richter scale ranking below 2 cause no damage and are not usually felt by people, but will be recorded by seismographs. Earthquakes that have a Richter scale ranking between 2 and 4.5 usually cause little or no damage and are felt by few people. Earthquakes that have a Richter scale ranking of 4.5 to 6 can cause minor damage and are usually felt by people. Earthquakes that have a Richter scale ranking from 6 to 8 can cause moderate to major damage and are felt strongly by people. Earthquakes that have a Richter scale ranking above 8 can cause extreme damage and are felt very strongly by people.

Although there is theoretically no upper limit to the Richter scale, there has never been an earthquake recorded that exceeded a ranking of 9 on the Richter scale.

Read the above passage again, paying careful attention to the details. Then answer the questions that refer to this paragraph.

Questions

1. A _____ is a scientist who studies earthquakes.

2. A _____ is an instrument that is used to measure the strength of an earthquake.

3. Each increase of one whole number on the Richter scale equals the release of _____ times more energy from an earthquake.

4. The point at which an earthquake originates is called the _____.

5. The strength and intensity of an earthquake _____ with distance away from the point of origin.

6. An earthquake with a Richter scale ranking of 5 would cause _____ damage to the area it affects.

7. An earthquake with a Richter scale ranking of _____ would cause extreme damage to the area it affects.

8. The northern part of New York state has many earthquakes with Richter scale rankings of less than 2. Would you expect there to be a great deal of earthquake damage in this area? Explain.

9. Scientists predict that Los Angeles will have an earthquake with a ranking of 7 to 8 on the Richter scale by 2020. Should Los Angeles prepare to deal with damage if such an earthquake occurred? Explain.

Answers

1. seismologist

2. seismograph

3. 31

4. epicenter

5. decrease

6. minor

7. 9 or higher

8. No, earthquakes of this magnitude are not strong enough to cause damage.

9. Yes, an earthquake of 7 to 8 on the Richter scale could cause major damage.

DRAWING CONCLUSIONS

Sometimes you have to reach beyond scientific information as it is presented and use your knowledge and skills to make a determination or draw a conclusion. Being able to draw a conclusion from a scientific concept, theory, or presentation is an important skill that you can use to solve scientific questions and problems. Drawing conclusions is often as simple as being able to put different pieces of information together. Knowing how different pieces of information relate to one another allows you to make a logical assumption, otherwise known as a conclusion.

Your local weather forecaster is constantly drawing conclusions from the current weather information in order to make an accurate forecast of future weather. For example, suppose the weather forecaster knows that an air mass with temperatures in the 40s is moving in tomorrow along with a storm. He or she can conclude that any precipitation that occurs as a result of the storm will be in the form of rain, since a temperature of 32 degrees Fahrenheit or less would be necessary for snow or sleet to occur.

Drawing a conclusion is by no means simply making a guess. You must have a firm understanding of the subject matter and base your conclusion on scientific principles. For example, you could say that it is going to snow next Christmas and base it upon the fact that it had snowed last Christmas. However, the prediction is a guess and not a conclusion, since it is not based upon scientific weather information, but rather is based upon an unsupportable and illogical assumption that since it snowed last Christmas, it will snow again this Christmas.

Take a moment to familiarize yourself with the weather map on the following page (Figure 12). A warm front (half-circles) and cold front (triangles) are depicted on the map. The symbols point in the direction that the front is moving. Snow occurs at temperatures below 32 degrees Fahrenheit. Rain occurs at temperatures above 32 degrees Fahrenheit.

Earth Science

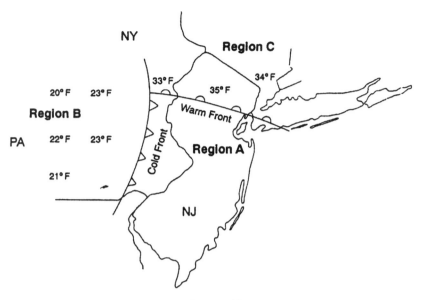

Figure 12

Questions

1. What type of precipitation would occur in Region B? _____

2. What type of precipitation would occur in Region C? _____

3. What temperatures would you expect to be present in Region A? _____

4. If the cold front continued to move in its present direction, what temperatures would you soon expect to be present in Region A? _____

Answers

1. Snow; snow occurs at temperatures below 32 degrees Fahrenheit, and the temperatures in Region B are below 32 degrees Fahrenheit.

2. Rain; rain occurs at temperatures above 32 degrees Fahrenheit.

3. Higher than 35°F; we know it is warm in Region A since a warm front is moving toward Region C.

4. 20–23° F; the cold front is moving from Region B, where the temperatures were between 20 to 23 degrees Fahrenheit.

Read the passage below carefully.

The burning of fossil fuels such as coal and oil releases carbon dioxide into the atmosphere. Since the dawn of the Industrial Revolution, man has been burning fossil fuels at an increasing rate, while carbon dioxide levels have been increasing at a similar rate. Carbon dioxide in the atmosphere traps sunlight that normally bounces off of the Earth and goes back into space. This is called the greenhouse effect. The majority of scientists believe that the result of increasing carbon dioxide levels will be an increase in the average global temperature, a theory called global warming. The Earth's atmosphere has warmed by approximately one degree Fahrenheit since the dawn of the Industrial Revolution. Many consecutive years within the last decade have broken the record for being the warmest ever recorded.

Concerned scientists point to these facts as proof that rising carbon dioxide levels are already resulting in a man-induced global warming. However, there is a minority of scientists who think that it is still too soon to determine if rising carbon dioxide levels are causing a man-induced global warming.

These scientists point out that we could be in a natural warming cycle that has nothing to do with man-induced global warming. Some scientists even suggest that if man-induced global warming is occurring, then nature could create a backlash that would cause the planet to cool dramatically. These scientists point out that nature tends to balance out extremes through unexpected means. In the case of global warming, a dramatic increase in atmospheric moisture and clouds caused by global warming could eventually have a global cooling effect on the Earth's atmosphere.

Questions

1. What conclusion can you make about scientific opinion concerning the issue of human-induced global warming?

   _____
   _____
   _____

2. What conclusion can you make about the burning of fossil fuels as it relates to carbon dioxide levels?

   _____
   _____
   _____

Answers

1. Scientific opinion regarding global warming is mixed. However, a majority of scientists believe it is occurring.

2. Burning fossil fuels increases carbon dioxide levels.

READING A CIRCLE GRAPH

A **circle graph**, also known as a **pie graph**, is a solid circle that is broken into wedge-shaped pieces with pointed ends at the center. The size of each wedge proportionally represents data. Circle graphs present data in a clear and easily understandable format. A glance at a circle graph can present information that otherwise could take more time to interpret. Circle graphs are also used to *enhance* a written statement. A point that is being made in written literature becomes more powerful when the data being discussed is also presented in a circle graph. Science writers often include circle graphs in their writings to enhance their presentation.

Circle graphs are usually used to present basic data that is obtained at a specific point in time or pertains to a specific subject matter. For example, a circle graph could be used to show how the public feels about protecting the environment on the day that the poll is taken. In order to show the results of another opinion poll on the same subject taken a year later, an entirely new circle graph would have to be constructed and placed beside the original circle graph in order to compare the change in public opinion from one year to the next.

Question

1. What are some other types of scientific information that could be presented clearly using a circle graph?

    _____

    _____

    _____

Answer

1. Types of storms that occur in a given year, intensity of earthquakes in a region in a given year, etc.

A recent survey of college students found that 75% (three-quarters) of the students think that protecting the environment should be a top priority, while 25% (one-quarter) think that protecting the environment should not be a top priority. A circle graph representing this information would have 75% (three-quarters) of the circle labeled as "college students who think that protecting the environment should be a top priority" and 25% (one-quarter) of the circle labeled as "college students who don't think that protecting the environment should be a top priority."

The first piece of information to look for when reading a circle graph is its title. The title will tell you what information the circle graph is presenting. The pieces of information depicted on a circle graph can either be written on the circle graph or can be contained in a graph key. Check to see if there is a graph key associated with the circle graph. The graph key will have the pieces of information depicted on the circle graph listed with a distinct color, shading, or pattern next to each type of information. Familiarize yourself with the graph key and then look at the circle graph. The pieces of information contained in the graph key will be presented on the circle graph in the distinct color, shading, or pattern in a way that is directly related to the data being presented. Take a look at how this information is presented in the circle graph and graph key on the following page (Figure 13).

Note that there are distinct shading patterns associated with the categories "Endangered Species," "Pollution Control," and "Recycling." These patterns are found on the circle graph in amounts proportional to the number of people who believed that a certain area of environmental protection should be a priority. A total of 300 people were surveyed in this poll. "Pollution Control" was listed as the top priority of 75 people surveyed. "Recycling" was listed as the top priority of 75 people surveyed. "Endangered Species" was listed as the top priority of 150 people surveyed. Since 75 is 25% of 300, 25% (one-quarter) of the circle graph is represented by the "Pollution Control" pattern and another 25% (one-quarter) of the circle graph is represented by the "Recycling" pattern. Since 150 is 50% of 300, 50% (one-half) of the circle graph is represented by the "Endangered Species" pattern.

As you can see, the amount of people who felt that pollution control should be the top priority is clearly represented as one-quarter of the circle graph. Even if you did not immediately know that 75 is one-quarter of 300, you can quickly and easily see this fact presented on the circle graph. This is a very simple example. The information presented in circle graphs in scientific literature is usually more complex. However, by looking at the size of the wedges in a circle graph you can quickly understand even complex information.

Take a moment to familiarize yourself with the two circle graphs shown on the fol-

GED Science

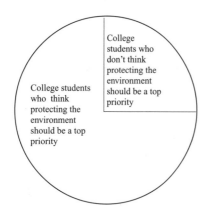

Results of a college student survey on protecting the environment

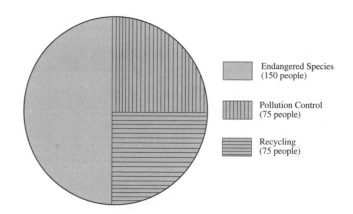

Public opinion of the top priority in environmental protection (300 people polled)

Figure 13

lowing page (Figure 14). These graphs present the amount of individual air pollutants that are found in smog-polluted air and the amount of air pollutants that come from various sources.

Questions

1. What type of air pollutant is the largest constituent of smog? _____

2. What is the largest source of air pollution? _____

3. What type of air pollutant is the smallest constituent of smog? _____

4. What source produces the smallest amount of air pollution? _____

5. What percentage of smog is caused by sulfur oxides and nitrogen oxides combined? _____

6. Based upon the information presented in the circle graph, which source of air pollution would you try to control in order to make the biggest impact on the problem of smog? _____

Answers

1. Carbon monoxide makes up 47 percent of the pollutants that are found in smog-polluted air.

2. The largest source of air pollution is transportation (42%).

3. The smallest constituent of smog is nitrogen oxide (10%).

4. Solid waste disposal produces the smallest amount of air pollution.

5. 25%; sulfur oxides constitute 15% of smog, nitrogen oxides constitute 10%.

   15% + 10% = 25%

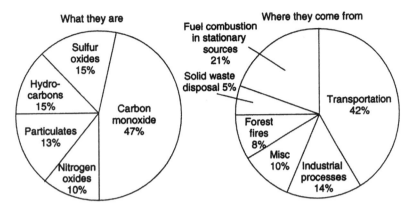

Figure 14

6. Since transportation is the largest source of air pollution, you need to try to control its large impact on the problem of smog.

Take a moment to familiarize yourself with the circle graph shown to the right (Figure 15). This graph depicts the amount of carbon dioxide that is present in Earth's oceans, atmosphere, geology (carbon dioxide that is contained in rocks and fossil fuels that make up the Earth's surface layer to a depth of a few miles), and terrestrial forms (carbon dioxide that is contained in living organisms including plants and animals). Carbon dioxide is released into the Earth's atmosphere by burning fossil fuels such as oil and coal. As carbon dioxide builds up in the Earth's atmosphere, it traps sunlight which then warms up the atmosphere. The warming of Earth's atmosphere by the increase of carbon dioxide levels is called global warming.

Questions

1. Where is the majority of the Earth's carbon dioxide? _____

2. How many tons of carbon dioxide are contained in the Earth's geology? _____

3. How much carbon dioxide is contained in the Earth's atmosphere? _____

4. If half of the carbon dioxide from the Earth's geology was transferred through the burning of fossil fuels into the Earth's atmosphere, how much carbon dioxide would be in the Earth's atmosphere? _____

Answers

1. oceans

2. $4{,}000 \times 10^9$ tons

3. 748×10^9 tons

4. $2{,}748 \times 10^9$ tons; there are $4{,}000 \times 10^9$ tons of carbon dioxide contained in the Earth's geology. Half of the Earth's geology would be

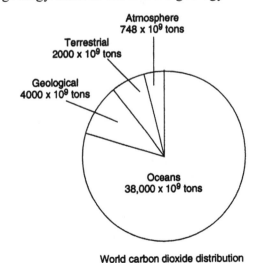

World carbon dioxide distribution

Figure 15

$2{,}000 \times 10^9$ tons. If half of the carbon dioxide from the Earth's geology was transferred through the burning of fossil fuels, then we would add $2{,}000 \times 10^9$ tons and 748×10^9 tons (carbon dioxide in the Earth's atmosphere). This is how we get $2{,}748 \times 10^9$ tons.

READING A BAR GRAPH

A **bar graph** is a graph that uses rectangle-shaped forms, called bars, to display information. Bar graphs are used to summarize information in an easily understood format. The use of a bar graph in scientific literature can greatly enhance the presentation of information. Bar graphs are especially useful for the presentation of complex information that covers a period of time or multiple subjects.

A bar graph has a horizontal axis (from left to right), called the X axis, and a vertical axis (from bottom to top), called the Y axis. The X and Y axes contain distinct information about a subject matter and are labeled separately. The X and Y axis labels also include the units of measurement of the information being presented.

For example, let's consider a bar graph that shows the number of hurricanes that have occurred in the Atlantic Ocean from 1966 through 1970 (Figure 16). The X axis (horizontal) lists the years 1966 through 1970 from left to right and is titled "Hurricane Seasons (Years)." The Y axis (vertical) lists whole numbers starting with zero from bottom to top and is labeled "Number of Hurricanes Per Season." The number of hurricanes from 1966 though 1970 is shown by a bar extending upwards above each year along the X axis to a height along the Y axis that equals the number hurricanes that occurred in that year.

Being able to see and interpret a trend, which can appear on a bar graph, is an important aspect of collected information. A trend is the general direction, drift, or tendency of a set of data, usually over a period of time. For example, if the number of Atlantic hurricanes increases by one hurricane each hurricane season for three years, the trend is an increase of hurricane activity over the period of those three years.

Questions

1. What year had the greatest number of hurricanes in the time period covered by the bar graph? _____

2. What year had the least number of hurricanes in the time period covered by the bar graph? _____

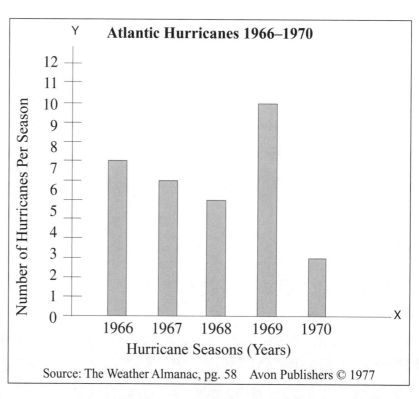

Figure 16

3. Can you see a trend in the number of hurricanes between 1966 and 1968?

4. Does this trend continue into the 1969 hurricane season?

5. How many more hurricanes occurred during the 1969 hurricane season as compared to the 1970 hurricane season?

6. What general observation can you make about the frequency of hurricane occurrences per hurricane season based upon the information provided by the bar graph?

Answers

1. 1969; there were ten hurricanes in 1969.

2. 1970; there were three hurricanes in 1970.

3. The trend is downward, less hurricanes; In 1966, there were seven hurricanes; in 1967, six; and in 1968, five.

4. No; in 1969, there were ten hurricanes, the greatest number of hurricanes from 1966 to 1970.

5. 7; there were ten hurricanes in 1969 and three in 1970.

    $10 - 3 = 7$

6. The frequency of hurricanes varies from year to year, but some minor trends can occur.

The bar graph on the following page (Figure 17) demonstrates how two pieces of information can be presented simultaneously in a graph. Read the title and graph key carefully before interpreting the bar graph.

The bar graph on the following page presents the total number of Atlantic hurricanes during the 1966 through 1970 hurricane seasons, along with the number of Atlantic hurricanes making landfall through the same period. Reading and understanding the title of the bar graph and the graph key is essential when you are trying to comprehend the information presented in a bar graph. The total number of Atlantic hurricanes and the number of Atlantic hurricanes that made landfall in the United States are distinctly represented by different patterns within the bars.

Questions

1. What observation can you make about the number of Atlantic hurricanes that made landfall in the United States during the 1966 through 1970 seasons compared to the total number of Atlantic hurricanes that formed during this period?

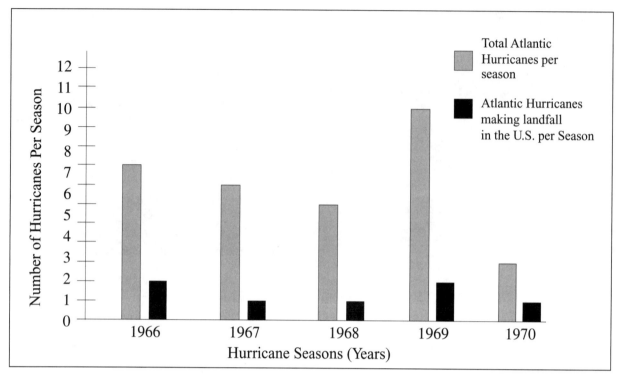

Figure 17

2. What years had the most Atlantic hurricanes that reached landfall in the United States? _____

3. Are there any similar trends concerning the total number of Atlantic hurricanes and the number of Atlantic hurricanes that made landfall in the United States during the 1966 through 1970 hurricane seasons?

   _____
   _____
   _____

4. Which hurricane season had the biggest difference between the total number of Atlantic hurricanes and the number of Atlantic hurricanes that made landfall in the United States? _____

Answers

1. There was no consistent correlation between the two statistics. Hurricane formation does not directly relate to hurricanes making landfall in the United States.

2. 1966, 1969; there were two in 1966 and 1969.

3. There is no apparent relation.

4. 1969; there were ten hurricanes in 1969 and only two made landfall.

Earth Science

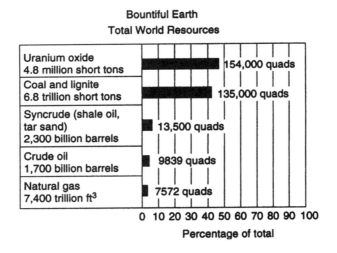

Figure 18

Take a moment to familiarize yourself with the bar graph shown above (Figure 18). This bar graph presents information concerning the total known world resources of the major types of fossil and nuclear energy sources.

Questions

1. How is the above bar graph different from the other bar graphs in this section?

2. What type of energy has the largest known reserve? _____

3. What percentage of the total does this energy source represent? _____

4. What type of energy source is 42% of the total? _____

5. Does this bar graph give you enough information to determine how long it would take to deplete all of the known reserves of natural gas? Explain.

Answers

1. It is horizontal, not vertical

2. Uranium oxide; it has 154,000 quads.

3. Approximately 45%

4. Coal and lignite

5. No, there is no depletion time information.

105

TOPICS IN EARTH SCIENCE

This material will provide you with an overview of the earth sciences field. These are topics you will need to become familiar with before taking the GED examination.

Study of Earth Science Defined

Earth science is the study of the Earth and its parts. Earth science has many subgroups, including, but not limited to, astronomy, geology, meteorology, and oceanography.

The general study of the earth sciences encompasses all scientific fields. The earth scientist strives to have a diverse background in all areas of science. Such a background includes the broad science areas of biology, chemistry, and physics. Because it is such a diverse science and draws on so many fields of study, the vocabulary, techniques, and concepts of earth science often cross the gray boundaries that we perceive to separate the sciences. An earth scientist needs to be familiar with all areas of the physical and life sciences and works to make connections between present findings and past research. The earth scientist works to better understand the present and the past history of Earth, as well as looking to the future of our planet.

Prediction

From meteorology to planetology, prediction is a cornerstone of science. What, when, where, why, and how are all questioning tools used to understand the natural world and universe around us. With a basic understanding of the elements of nature, science allows for predictions of future events based on past observations and evidence. The weather for a particular area can be hypothesized based on past climate and weather. Geologists can estimate time frames for volcanic eruptions using data from past events such as seismic activity, slope bulge increases, gas volumes, and concentrations of gases emitted from volcanic vents. The past is a key to today, but perhaps the past is an even greater tool in science for the prediction of events that will occur in the future.

The Branches of Science Within Earth Science

The study of earth sciences includes all aspects of the physical earth. This includes the oceans and atmosphere, mountains and valleys, volcanoes and earthquakes, and the evolution of earth systems—past, present, and future. The rocks and waters of Earth are studied by geologists as well as the minerals and resources that support our civilizations. The earth sciences also include research in areas far beyond the sphere of our tiny planet. Planetologists speculate and study about far-off worlds and moons. Astronomers peer into deep space to gather evidence of the origins of the universe and celestial bodies. They help us to look to future exploration with anticipation and wonder. Lessons learned here can be applied to far-off worlds to gain insight to cosmologic events in space and how they may relate to Earth.

Living on a planet whose surface is covered by 70 percent water, the oceanographer has an extensive role in developing an understanding of the connections between Earth and the oceans. The number of areas to be considered is so vast that the field of study can include formation of shorelines, the features of the sea floor, and the processes that caused the formation of the ocean basins. Additional areas of study include tides and currents, air-water interaction, water chemistry, heat transfer, weather and climatic effects and interactions, organisms, environments, and resources. The topics listed are not limited to the domain of oceanography, but are interconnected and thus important to a comprehensive study of oceans.

Even if we try to ignore the weather, we are likely to fail, as meteorology will in some way directly impact our lives. Traditionally, this area of the earth sciences is divided into two disciplines. Meteorology is concerned with the daily short-term fluctuations in atmospheric phenomena. The composition and structure of the atmosphere, incoming solar radiation, ocean interaction with air, evaporation, heat transfer, cloud formation, and precipitation are all topics of interest within this area of study. Climatology is defined as the long-term effect of weather on a given area. The causes of climate and climatic variation are of primary interest in this area of study. The long-term effects of weather can be better understood by connecting and understanding global system interactions. The past and present climates of Earth are areas of interest to meteorologists, as well as predictions of possible future climatic changes.

A deep understanding of the planet Earth is needed to explain its natural aspects. The primary goal of the geologist is to understand the changes that take place on and below the surface of the planet, and to understand the reasons and causes of those changes. Physical geology is the study of Earth's materials, surface changes, internal structure, and the forces that cause those changes. The geologist strives to know structures and form, as well as the historical past and possible future events of Earth. Wind and water are processes that contribute to the shaping of Earth and, thus, are important factors to understand. Resources and reserves are areas germane to geology as well.

To contemplate the complete past of the universe is indeed a grand endeavor. The knowledge gained in the study of the past can help space scientists to better understand cosmology in the present and perhaps predict events and changes in the future of the universe. Closer to home is the study and understanding of planets and moons, asteroids, comets, and meteorites. Perhaps the greatest contributions of these researchers to the knowledge of Earth is the increasing body of information available on terrestrial (earth-like) and Jovian (giant gas) planets, e.g. Jupiter. The comparative geology allowed due to the huge mass of data for comparison has increased awareness of Earth's features and internal and external processes that cause those features.

Physics can be simply defined as the study of forces, motion, energy, and their effects on matter. The pure physics of a scientific laboratory transfers easily to the physical world called Earth. Crustal motion, earthquakes and volcanoes, atmospheric and oceanographic motions, heat flow in magma and water, and solar output and electromagnetic spectrum are just a few of the many areas where physics plays a part in the earth sciences. Few scientific endeavors are confined within the bounds of a "pure" science. Science and the earth sciences in particular are versatile in nature. That is, input from many sources and disciplines is needed to provide the whole big picture of Earth.

Chemistry is the study of the properties and composition of matter in the natural world. Matter is defined as the physical material of the universe, which occupies space and has mass. The Earth and all the geological features we see are the result of physical and chemical changes. The core of the Earth, through radioactive decay, provides heat to generate gigantic convection cells. Those cells in turn manipulate the crust of the Earth as huge crustal plates collide. The pressures and temperatures produced cause magma to form. Those magmatic bodies affect the surface of Earth as geologic landforms are produced and precious mineral resources are deposited. Not only are crustal processes of importance to a chemist, but those of the atmosphere and hydrosphere are of interest as well. The chemistries of the oceans and air are also germane to the chemist. Again, the earth sciences prove to be versatile in nature as many disciplines contribute to the big picture of understanding the nature of Earth.

Defined as the study of living organisms, biology proves to be another valuable tool to assist in the development of a comprehensive understanding of the earth sciences. The most obvious application in geology would be the study of paleobiology, or the fossil record. Another application of biology within the realm of earth science is in the understanding of the deposition, formation, and location of fossil fuel resources. In areas of resource management and pollution, the ecological and environmental aspects of biology prove to be critical. Knowledge of ocean life can be combined with geologic and oceanographic information to provide a more complete picture of the complex environment we call the ocean.

Once again it must be stated that the sciences are all interconnected in a complex web of knowledge. The boundaries that define disciplinary borders become gray with the passage of time and the accumulation of information. The sciences are all intertwined both in purpose and substance as the universe and Earth become better known and understood.

Volcanology—The **volcanologist** needs to draw from many specialized fields of study to undertake a complete study of volcanic formation and eruptions. Knowledge of chemistry will allow for an understanding of magmatic composition and the releases of gas from the magma. Subtle changes in the composition of gases may hint at future changes in store for any particular volcano. Basic geology will provide some historical background. A knowledge of physics and mathematics will prove to be very useful to model past and present data into a workable hypothesis of future events. Computer analysis is of paramount importance in this field, as in any area of science. The mass of data being accumulated has to be quickly and accurately analyzed so that assumptions and predictions can be made about future volcanic events.

Glaciology—Knowledge of basic geology is very important to a glacial expert. The landforms, lakes, and general topography of a region are tell-tale signs of past glacial events. An understanding of climatology and meteorology is essential for understanding past fluctuations in glacial surges and retreats. Global fluctuation in climate is responsible for changes in the total percentage of ice covering Earth today. To predict future changes in glacial masses, models need to be conceived and adapted. Any subtle change in sea level would affect millions of people. A rise in ocean waters could inundate coastal communities, while a falling sea caused by glacial growth and advance could leave important ports high and dry. These changes in glacial coverage could also have dramatic impact on crop-growing belts and world agriculture. The correlation between glaciers, climate, and humans is a most important field of study for many of the reasons listed above.

Seismology—The surface of Earth is an ever-changing place. The shifting of the crust directly affects millions each year. Recent examples of tectonic activity in Japan and the United States points out the importance of a thorough understanding of crustal movement. Earthquakes occur all over the surface of Earth and have also been recorded on other bodies in our solar system. The potential for economic and human disaster is enormous. The major goals in understanding earthquakes are the prediction of three important pieces of information. The prediction would include the potential place an earthquake would occur, the predicted magnitude or size of the quake, and the time frame in which the shaking could be expected. Knowledge of all areas of the sciences is crucial to fully understand this most complex and potentially lethal reaction created by the enormous stresses that occur within the Earth.

Marine Geology—The marine geologist is sometimes called the geological oceanographer. This scientist uses the tools and knowledge of both fields to further understanding of the sea floor, landforms, and their formation. An understanding of tectonics assists this type of science specialist. Also needed is knowledge of tides, currents, and geomorphology of sea-floor features. Knowledge of paleoclimates helps in the understanding of sea-level fluctuations and changes in coastal topography. The use of computers and modeling also assists the marine geologist, as the volumes of new data to be assimilated and understood is massive.

Planetology—The study of our solar system begins with an extensive knowledge of our planet, Earth. We have found many similarities between Earth and our surrounding neighbors. The composition of other moons and planets has provided insight into the processes that formed our planet. Rivers and erosion, glaciers, volcanoes, and tectonics are some of the phenomena observed in our exploration of the solar system. The staggering amount of data collected in recent decades is still being processed and reevaluated. Ideas are formulated and hypotheses changed due to new input and understanding. The possible visual and mathematical observation of a new solar system beyond our own has broken open the future possibilities for planetologists.

Climatology—**Weather** can be defined as the short-term changes in the atmospheric conditions near the surface of Earth. **Climate**, on the other hand, consists of the long-term conditions that affect land masses, plants, animals, and humans. Rainfall and temperature are two of the components of the study of climatology. Past and future trends in climate are a concern of this science. Modeling and computers are fundamental in working with systems and trends within the atmosphere of Earth. The air and its interaction with oceans and land masses increase our understanding of complex patterns and systems. The influx of solar energy is also of primary importance in the study of climatology. Knowledge gained can be applied to atmospheric systems on other planets and moons as well. As more information is gained and understood, greater application of climatic models can be made for Earth and all bodies in the solar system.

Astronomy

Astronomy is the study of celestial bodies and their movements.

The Earth is one of nine planets in our solar system. A solar system is composed of a star and the objects that move about it. The largest objects moving about a star are called planets. The planets in our solar system, beginning at the sun and moving away from it, are Mercury, Venus, Earth, Mars, Jupiter, Saturn, Uranus, Neptune, and Pluto.

Many objects smaller than planets exist in our solar system. If one of these smaller objects collides with Earth, it is called a **meteor** or "shooting star." The glow of a meteor is caused by its burning as it passes through our atmosphere. Meteors that reach the Earth's surface are called **meteorites**.

The Earth's path around the sun is called its **orbit**. The Earth's orbit around the sun, called a **revolution**, is completed in $365\frac{1}{4}$ days. An axis is an imaginary line passing through the poles of the Earth. The Earth spins on its axis, and each spin is called a **rotation**. One rotation takes 24 hours. Rotation causes the alternation of day and night on Earth.

The Earth revolves about the sun. Its axis is tilted $23\frac{1}{2}°$ perpendicular to the plane of Earth's orbit (See Figure 19). This tilt results in differing proportions of day and night on Earth throughout the year, and also causes the seasons. Day and night are of equal length only twice each year, at the **autumnal equinox** and **vernal equinox** (the first day of autumn and the first day of spring).

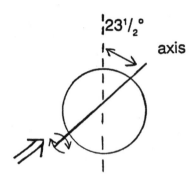

Figure 19

The moon is a satellite of Earth. It moves in orbit about Earth, and one revolution takes $27\frac{1}{3}$ days. The moon reflects sunlight, which causes it to glow. When the Earth blocks sunlight from reaching the moon, it creates a shadow on the moon's surface, known as a **lunar eclipse**. If the moon blocks sunlight from hitting the Earth, a **solar eclipse** is created. The moon has a gravitational pull on Earth which causes **tides**, or periodic changes in the depth of the ocean.

Geology

Geology is the study of the structure and composition of the Earth.

The Earth is composed of three layers—the crust, mantle, and core (See Figure 20). The **core** is the center of the Earth, and is made of solid iron and nickel. It is about 7,000 km in diameter. The **mantle** is the semi-molten layer between the crust and core, and is about 3,000 km thick. The **crust** is the solid outermost layer of the Earth, ranging from 5–40 km thick. It is composed of bedrock overlaid with mineral and/or organic sediment (soil).

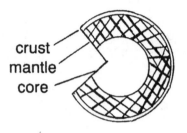

Figure 20: Layers of the Earth

Large sections of the Earth's crust, called **plates**, move at times, creating earthquakes, volcanoes, faults, and mountains. The study of these movements is called **plate tectonics**. **Faults** are cracks in the crust formed when plates move. Faults gape open when plates move apart and are closed when plates slide past one another. Earthquakes occur when plates slide past one another abruptly. Earthquakes may also be caused by volcanoes. Earthquakes are measured by a seismograph on the Richter scale.

Volcanoes form where plates move away from one another to let magma reach the crust's surface. **Magma** is molten rock beneath the Earth's crust. Lava is molten rock on the Earth's surface. Mountains are formed by volcanic activity or the collision of plates, which causes the crust to buckle upward.

Rocks are naturally occurring solids found on or below the surface of the Earth. Rocks are made of one or more minerals. Minerals are pure substances made of just one element or chemical compound. Rocks are divided into three groups, based on the way they are formed.

1. **Igneous**—rocks formed by cooling of magma or lava (e.g. granite, obsidian).

2. **Sedimentary**—rocks formed from silt or deposited rock fragments by compaction

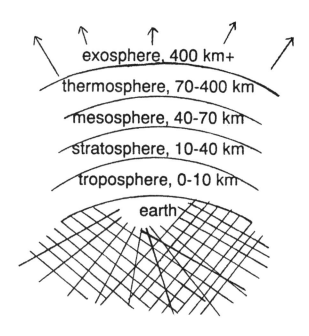

Figure 21: The Layers of the Atmosphere

at high pressures and/or cementation (e.g. shale, limestone).

3. **Metamorphic**—rocks formed from igneous or sedimentary rock after exposure to high heat and pressure (e.g. marble, slate).

Weathering is the breaking down of rock into small pieces. Rock is weathered by acid rain, freezing, wind abrasion, glacier scouring, and running water. Erosion is the transportation of rock or sediment to new areas. Agents of erosion include wind, running water, and glaciers.

Meteorology

Meteorology is the study of the atmosphere and its changes.

The **atmosphere** is a layer of air surrounding the Earth. Air is a mixture of gases, the most common being nitrogen and oxygen. The atmosphere is studied because 1) it protects (insulates) the Earth from extreme temperature changes, 2) it protects the Earth's surface from meteors, and 3) it is the origin of weather.

The atmosphere can be divided into several layers (See Figure 21). The **troposphere** is the layer closest to Earth. Almost all life and most weather is found there. The **stratosphere** is the chief thermally insulating layer of the atmosphere. It contains the ozone layer and jet stream. The stratosphere is the region where ozone is produced. The **thermosphere** causes meteors to burn up by friction as they pass through. This layer reflects radio waves. The **exosphere** is the outer layer of the atmosphere. It eventually blends into the vast region we call "space."

Weather is the local, short-term condition of the atmosphere. The two factors that affect weather most are the amounts of energy and water present. Most of the energy that affects weather comes from the sun. As solar energy hits the Earth, most of it is scattered or reflected by the atmosphere. The solar energy that gets through the atmosphere warms the Earth's crust, which in turn warms the atmosphere. The Earth does not absorb solar energy uniformly; the equator absorbs more than the poles do. This difference in energy (heat) absorption, in part, causes winds.

Water covers about 70 percent of the Earth's surface. As that water slowly evaporates, some of the vapor is held in the atmosphere. It is the water vapor in our atmosphere that causes humidity, fog, clouds, and precipitation.

An **air mass** is a huge area of air that has nearly uniform conditions of temperature and moisture. When two air masses meet, the boundary between them is called a **front**. Fronts are the location of most stormy weather.

Warm air is less dense than cold air. That means a given volume of warm air weighs less than an equal volume of cold air. Air masses push down on the Earth below them, causing air pressure. Warm air masses, because they are less dense, push down less and cause low pressure areas. Cold air masses, which are more dense, cause high air pressure. Air moves from high pressure to low pressure areas, causing wind.

Different aspects of the weather may be measured using special instruments.

Weather Aspect	Instrument
wind speed	anemometer
air pressure	barometer
humidity	hygrometer
temperature	thermometer

Clouds can be used to predict the weather.

Cloud Type	Appearance	Weather
stratus	flat, broad	light colored—stable weather conditions
		dark colored—rain expected soon
cumulus	fluffy, solid-looking	light colored—good weather dark colored—heavy rains, perhaps thunderstorms
cirrus	thin, wispy	changes in weather

Climate is the general atmospheric condition of a region over a long period of time.

Oceanography

Oceanography is the scientific study of the ocean and marine life, including its biology, chemistry, and physics.

Sea water differs from fresh water in its salinity, or saltiness. Fresh water, the water we drink, has relatively few dissolved solids in it and has low salinity. Ocean water has a lot of dissolved material in it and therefore has a high salinity. Many materials are dissolved in sea water, but the most abundant dissolved material is common salt, sodium chloride.

Ocean waters move through tides, waves, and currents. **Tides** are periodic changes in ocean depth. They are caused by the gravitational pull of the moon on Earth. Most waves are caused by winds. Some ocean currents are caused by density differences in sea water. Currents are like rivers within the ocean. The swift moving water in currents can transport material over large distances very quickly.

Practice: Earth Science

DIRECTIONS: Use the maps and passages below to answer the section review questions.

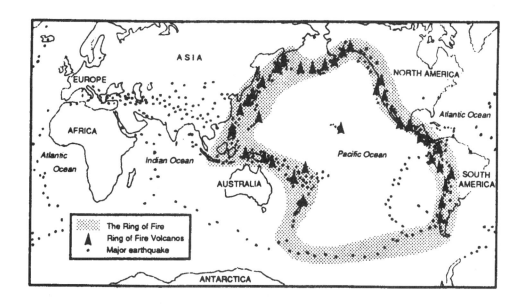

1. Circle the title that best fits the above map.

   (A) "World Continents and Oceans"

   (B) "Major Earthquake Locations"

   (C) "The Ring of Fire"

   Circle "T" if the statement is true; and circle "F" if the statement is false.

2. The Ring of Fire is located in the Atlantic Ocean.

   T   F

3. Major earthquakes occur within the Ring of Fire.

   T   F

   Use the topographic map shown on the following page to answer the following questions. Circle "T" if the statement is true, and circle "F" if the statement is false.

4. There are at least two gaging stations on the river cutting across the middle of the map.

   T   F

5. Natural Arch is at an elevation of approximately 4,400 feet.

   T   F

   Fill in the blanks.

6. Cheops Pyramid has an elevation of _____ feet.

7. _____ is the closest landmark to Buddha Cloister.

   Carefully read the paragraphs below.

   El Niño is a phenomenon in the Pacific Ocean in which the surface water temperature of a portion of the ocean rises by about eight degrees Fahrenheit. The west-

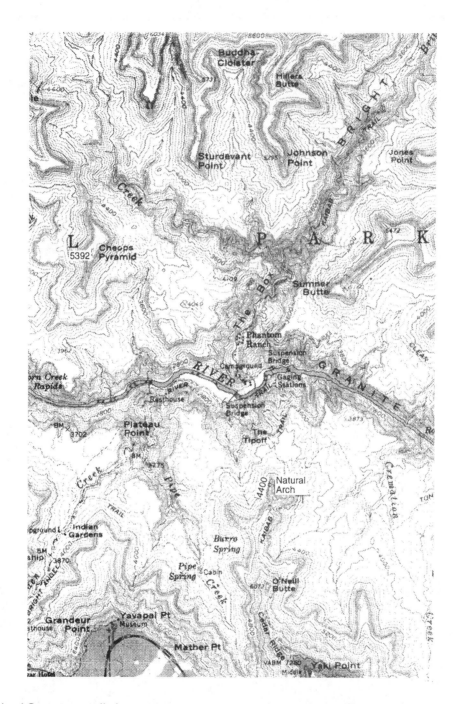

ern United States usually has very stormy winter seasons with flooding rains when El Niño is occurring. The eastern United States often experiences unusually mild or unusually cold winters when El Niño is occurring.

Weather in the United States is affected dramatically by the jet stream, a high-speed wind current in the upper atmosphere. The jet stream guides storms and air masses around the Earth. Scientists believe that there is a connection between the temperature of the oceans and the position of the constantly shifting jet stream.

8. Based upon the paragraphs above, choose the conclusion that makes the most sense.

(A) The jet stream is not affected by anything in the natural world.

(B) A stormy winter season in the western United States causes a mild winter in the eastern United States.

(C) El Niño affects the jet stream and causes storms to be directed by the jet stream into the western United States.

Circle "T" if the conclusion is true, and circle "F" if the conclusion is false.

9. If it is known in the fall that El Niño is occurring in the Pacific Ocean, people in the western United States should expect a stormy winter season.

   T   F

10. If it is known in the fall that El Niño is occurring in the Pacific Ocean, people in the eastern United States can expect a normal winter.

    T   F

11. Use the information provided in the following paragraph to construct a circle graph in the space provided. Give the circle graph a title.

    The nations of the world came together in Rio de Janeiro, Brazil in 1992 for an Earth Summit Conference to discuss global environmental problems. An agreement was reached at the conference to limit the release of carbon dioxide in order to control global warming. The United States releases approximately 25% of the total carbon dioxide released each year around the world. The other industrialized nations, combined, release approximately 25% of the carbon dioxide released each year. The remaining 50% of the carbon dioxide released each year comes from developing countries.

Draw Circle Graph Above

Use the bar graph below to complete the section review questions.

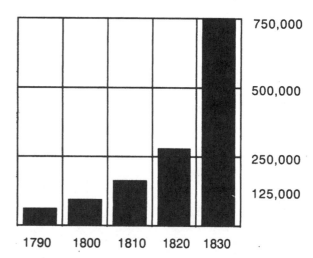

Bales of Cotton—1790-1830

Circle "T" if the conclusion is true, and circle "F" if the conclusion is false.

12. "Bales of Cotton" are represented by the Y axis.

    T   F

13. The number of bales of cotton produced in the 1820s was 500,000.

    T   F

Fill in the blank with the appropriate word.

14. There is a clear _____ of increasing cotton production in the period covered by this bar graph.

15. The number of bales of cotton produced in the 1830s was _____.

16. Circle the answer that best describes a compound word.

   (A) A compound word is a word that is associated with chemical compounds.

   (B) A compound word is a word that is an abbreviated form of a long word.

   (C) A compound word is a word that is comprised of two smaller words.

Circle "T" if the conclusion is true, and circle "F" if the conclusion is false.

17. Shower is a compound word.

   T   F

18. Snowstorm is a compound word.

   T   F

Answers

1. (C) Look at the map key. You will notice "The Ring of Fire."

2. False. The Ring of Fire is located in the Pacific Ocean.

3. True. Major earthquakes occur within the Ring of Fire.

4. True. There are at least two gaging stations on the river cutting across the middle of the map. ("Stations" is also plural, indicating more than one station.)

5. True. If you look at the map where it is labeled "Natural Arch," on the left of the label it says the elevation is 4,400 feet.

6. If you look on the map to the left of where it is labeled Cheops Pyramid, the elevation reads 5,392 feet.

7. Hillers Butte is located directly below Buddha Cloister on the left.

8. (C) Choice (A) is incorrect because scientists believe the temperatures of the ocean affect jet streams. Choice (B) is incorrect as an El Niño causes a milder winter in the eastern United States, not a stormy winter season in the western United States.

9. True. The western United States usually has very stormy winter seasons with flooding rains when El Niño is occurring.

10. False. When El Niño is occurring in the Pacific Ocean, people in the eastern United States often experience an unusually mild or unusually cold winter, not a normal winter.

11. Your circle graph should look like the one below. An example title for the graph could be "Release of Carbon Dioxide."

Release of Carbon Dioxide

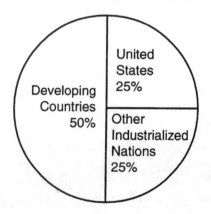

12. True. The numbers on the Y-axis represent the amount of bales of cotton.

13. False. The numbers of bales of cotton produced in the 1820s is a little over 250,000.

14. trend

15. 750,000

16. (C) Choice (A) is incorrect, because compound words do not have to be associated only with chemical compounds. Also, a compound word is not an abbreviated form of a long word (B).

17. False. Shower is not comprised of two smaller words.

18. True. Snowstorm is comprised of two smaller words: snow + storm.

REVIEW

In this section, we learned some very valuable skills for approaching material in earth science. These skills are reading a map, understanding compound words, finding details, reading a circle graph, and reading a bar graph.

Science

Physics

SCIENCE

PHYSICS

RECOGNIZING CAUSE AND EFFECT

When an object is pushed or pulled to make it move, we say that the push or pull is the cause of an effect—namely, the movement of the object. Turning a doorknob releases its bolt and the door is free to open. Here, the cause is the turning and the effect is the release of the bolt. Two operations of this kind are said to have a cause-and-effect relationship to one another. In a sense, the whole sequence of events that occurs in the world is a cause/effect chain, and where it is divided into these two categories is up to the person studying the situation. But one thing is always true—cause must come before effect; experience allows us to keep them correctly determined in the one-way direction of time.

There are certain key words and phrases to be aware of when determining what might be a cause and what might be an effect. Some of these phrases are "because of," "due to," "leading to," and "resulting in," besides single words like "thus," "makes," "causes" and "effects." The last two words are not nouns (representing a thing) but rather verbs (an action word) meaning "brings about."

Physics is the study of matter and energy moving about in space and time. One of the most basic ideas in physics is that of a machine. There are seven kinds of simple machines: the lever, wheel and axle, pulley, inclined plane, screw, gear drive, and the hydraulic pump. Any machine which includes two or more simple machines is called a compound machine.

Question

1. Name three compound machines you know about.

   _____

   _____

   _____

Answer

1. bicycle

   automobile

   washing machine

Read the following paragraph and answer the question that follows.

When an automobile's engine is on and you push the gas pedal, you are allowing more fuel to flow to the engine. This lets the engine run faster and, if the car is moving, it will go faster along the ground. The brake pedal has the opposite effect, so that pushing on it leads to the car slowing down. Turning the steering wheel to the right causes its axle to turn right which leads to the entire moving car turning to the right. Similarly, a left turn of the car is caused by turning the steering wheel to the left.

Question

1. In the above paragraph, circle key words and phrases relating to cause-and-effect. List some of the causes and effects you have found.

   Causes:

   _____

   _____

   _____

   Effects:

   _____

   _____

   _____

Answer

1. Causes:

   a) Turning steering wheel to the right

   b) Stepping on the brake pedal

   c) Pushing on gas pedal

   Effects:

   a) Automobile turns to the right

   b) Less fuel flows to the engine

   c) More fuel flows from gas tank into engine

Read the following paragraph and answer the question that follows.

A gear is a wheel with evenly spaced bumps on its outer edge. These bumps are called the teeth of the gear. An axle rod is attached to the center of the gear wheel. Two such gears may have their teeth meshed together so that the tooth of one will fill a depression between the teeth of the other. When the axle of one gear is turned, the meshed teeth cause the other gear to turn in the opposite direction. This system, in which one gear turns another, is called a gear drive. The two gears may also be separated but connected by a continuous drive chain. Here, one gear turning causes the drive chain to move and its motion is transferred to the other connected gear some distance away. In this case, the two gears are called chain gears.

Question

1. Circle the key words above relating to cause-and-effect. List several causes and effects from the information about gears.

Causes:

_____

_____

Effects:

_____

_____

Answer

1. Causes:

   a) Turning the axle on one gear that is connected to another gear

   b) Turning one gear that is connected by a drive chain to another gear

   Effects:

   a) The turning of a gear in the opposite direction

   b) The turning of a gear in the same direction

Read the following paragraph and answer the question that follows.

> The lever is a simple machine that caused the famous Greek scientist Archimedes (287–212 B.C.) to say, "Give me a place to stand and a lever long enough, and I will move the Earth itself!" What he meant by this is that a rigid rod, with one end placed under an object and with the rod resting on a rigid point along its underside, can be pushed down on its free end to produce a lifting of the object. The closer the rigid fixed pivot point, called the fulcrum, is to the object being lifted and the farther away the free end, the greater will be the mechanical advantage of the lever; it takes less push on the free end to lift a heavy weight.

Question

1. Circle any cause-and-effect words in the previous paragraph. Write some causes and effects about levers.

   Causes:

   _____

   _____

   Effects:

   _____

   _____

Answer

1. Causes:

   Pushing down on the free end of a lever pivoted on a fulcrum with its other end under a load

   Effects:

   The load is lifted into the air.

> Archimedes not only studied and used the lever a great deal, he also invented another one of the simple machines called the water screw. This machine was created in order to raise a liquid, like water, from one height to another through rotational motion. The screw is a long tube with a spiraled trough running continuously around its surface from one end to the other. At the upper level's end is a device—a person or a motor—turning the cylinder around its long axis while the other end is sitting in the liquid to be raised. Rotating the spiral trough causes the liquid to be carried upward toward the higher level, where it is drained off and used as needed.
>
> Many of the devices which Archimedes developed were used in warfare.

His lever principles were used to design catapults to hurl rocks and flaming projectiles at an enemy.

Question

1. Circle key cause-and-effect words in this last paragraph. Give an example of a cause-and-effect relationship in the lines below.

   Causes:

   _____

   _____

   Effects:

   _____

   _____

Answer

1. Causes:

   Turning a water screw

   Effects:

   Raises a liquid from one level to another

MAKING INFERENCES

The word **infer** means "to determine that something follows logically from something else." It may be the meaning of a single word or perhaps the truth of a general idea that is inferred.

When one object hits another object, scientists say that a collision has occurred between them. The object being struck may or may not be moving itself. The momentum of an object is the combination of its speed multiplied by how much matter it contains. If the total amount of momentum before collision equals the total amount of momentum after collision, scientists say the collision was elastic. Momentum can never be greater, overall, after collision than before. In real life, heat is always produced somehow between the two objects and this serves to reduce the overall momentum of any colliding system of objects.

Momentum is closely related to the energy of motion (kinetic energy of the objects in a moving system). The heat generated in a collision was not there before the two objects hit, but counting it into the situation at the end, let us say that the total amount of energy before collision equals the total amount of energy after collision. This fact is known as the Law of the Conservation of Energy. However, it is only true for closed systems in which the heat remains to be counted.

What can we infer from the preceding paragraphs? One thing is that an object that is not moving has zero momentum. Another might be that, in an inelastic collision, at least one of the objects gets hotter than before and maybe both do. A third inference might be that the Law of the Conservation of Energy is not true in open systems—in which heat escapes—because this is the logical opposite of the case referred to in the information given. This is a case of a logical inference being drawn.

Suppose someone catches a baseball using a thinly-padded mitt. If the baseball has a speed of 90 miles-per-hour, you could infer that it has a lot of momentum before being caught and zero momentum after. This means that the catcher's hand will hurt very much after stopping the ball since the mitt is poorly padded. The catcher's arm and hand must absorb the shock and have all of the ball's momentum transferred to them—except for a lot of heat that is produced in the mitt and the ball upon collision. This is why a catcher's mitt in regulation games has such thick and heavy padding in it.

One thing we could infer from this is that the speed of a caught baseball determines how much it will hurt with a thin mitt. The lower the speed, the less it will hurt; the greater the speed, the more it will hurt. Something else we could infer is that it is indeed possible for someone to throw a baseball at a speed of 70 miles-per-hour, or 80 miles-per-hour. This is because it is logical to assume that if a person throws a baseball at 90 miles-per-hour, they can throw it at any speed less than 90 miles-per-hour.

Electricity is the flow of special kinds of particles called electrons. The number of electrons moving either through a wire or in free space is called the **amperage** of the electric current. There is a condition that can make the electrons flow, similar to a push, and the measure of this electrical push is called **voltage**. When the direction of current flow is one way, scientists say it is a **direct current**. If the electrons shake back-and-forth, it is called an **alternating current**.

Read the following paragraph and answer the question that follows.

> When current flows, it usually encounters some resistance. In the nineteenth century, a German physicist named Georg Ohm discovered the law of direct currents that bears his name, "Ohm's Law." This law states that the amount of current through a continuous path called a circuit is greater if the voltage is greater, but lessened if the resistance is increased. That is, the greater the voltage the greater the amperage; the greater the resistance, the less the amperage. From this, we might logically infer that an alternating current requires that the voltage must be alternating, too, and this would be correct.

Question

1. Based upon Ohm's Law, what can you infer about the amperage in a direct current circuit if both voltage and resistance are increased or decreased by exactly the same amount?

   _____
   _____
   _____

Answer

1. The current stays the same.

Other words that mean infer are "deduce" and "imply." Whenever you see them, you are being asked to draw an inference. Some scientists prefer the word "deduction" to inference. However, they mean the same thing and presume that logic is being used to arrive at the inferred conclusion.

Read the following paragraph and answer the questions that follow.

> When an electrical current alternates, it gives off waves into space around the current carrier. These waves, called electromagnetic radiation, travel vast distances. The rate at which the current vibrates back and forth is called its frequency. The electromagnetic radiation will have the same frequency as the current. If the frequency of the radiation is about 1,000 trillion vibrations each second, then our eyes will detect it as visible light. The different shades of color we see are all just different frequencies of electromagnetic radiation in a fairly narrow band. No matter how the electromagnetic radiation is generated, or what its frequency is, it will always travel in free space with the same speed of 186,282 miles-per-second. Its direction of motion is straight unless it encounters an object or passes by a large planet. In that case, the radiation will be bent by the planet's gravity field.

GED Science

Questions

1. Cite three inferences you can draw from the given information on electromagnetic radiation.

   _____

   _____

   _____

2. Match the items in Column A with those in Column B.

A		B	
(A)	visible light	1.	decreases current flow
(B)	momentum	2.	Ohm's Law
(C)	alternating current	3.	depends on speed
(D)	resistance	4.	1,000 trillion cycles-per-sec.
(E)	direct current	5.	varying voltage in a circuit

Answers

1. All the colors we see are electromagnetic radiation, just less than or greater than 1,000 trillion vibrations-per-second. Light emitted by a fast object moves at 186,282 miles-per-second, which is the same as the source at rest. A star far behind a large cosmic object will not appear where it is if the large object is not there.

2. (A) 4
   (B) 3
   (C) 5
   (D) 1
   (E) 2

READING A DIAGRAM

A diagram is a line drawing that represents a system, an area on the ground or in space, or the relationship of two or more pieces of information. If the diagram is drawn so that one unit of distance, like an inch, on the paper shows some other unit of distance in the real world, then we say that the diagram is drawn to scale. When this is so, it is necessary to provide a scale key along with the diagram telling this relation of distance in the picture with distance on the large scale.

For example, look at Figure 1 shown below. This diagram is not drawn to scale because direct measurement with a ruler shows that the

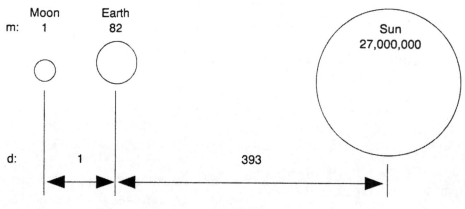

Figure 1

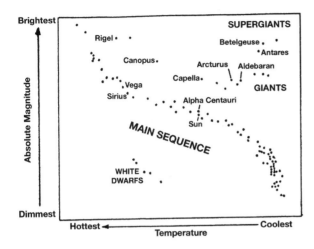

Figure 2

Earth–Moon distance pictured is about two centimeters. The diagram also states that there are 393 Earth–Moon distances from the Earth to the sun. However, if the diagram were drawn to scale, a picture 393 x 2 cm long (786 cm) between the circles showing the Earth and the sun would be needed! That is more than 25 feet of distance. So, this diagram is obviously not to scale.

The diagram above (Figure 2) shows the relationship between two pieces of information. Here, we are talking about the brightness and the temperature of the major stars.

Notice that most stars on this diagram fall along a curving middle path called the main sequence. The temperature gets hotter from right to left, so dots representing stars more to the left are hotter than those to the right. For example, the star Sirius is hotter than our sun because the dot representing it is to the left of the dot representing the sun. At the same time, the dot representing the sun is lower than the dot representing Sirius, and lower in this diagram means "dimmer" than the dots higher up. Therefore, the sun must be dimmer than the star Sirius.

There are many stars similar to our sun, as can be seen from the many dots near to it on the Main Sequence. However, there are many stars unlike it, too. Notice also that this diagram has absolutely nothing to say about the distance between stars. It only connects the two ideas of temperature and brightness together. Besides that, there is no indication of the positions of the stars in the night sky. We can't use it to go outside and locate a star from where we are standing at the time.

The diagram below (Figure 3) shows a small swinging object on a string. This system is called a simple pendulum. The arc of swing must not be too large and the string must be very light and thin for it to be called a "simple" pendulum. The object attached to the non-extendible string is shown by the circle at the end of the string (the line segment). A crosshatched area at the top of the diagram is meant to be some steady support like a ceiling where the simple pendulum is attached. A dotted line represents the position the string and ball (called "bob") have when they are not swinging. It is known as a vertical (i.e., straight up-and-down) dotted line. The position of the simple pendulum is shown measured from the

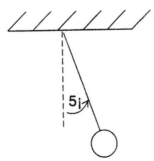

Figure 3

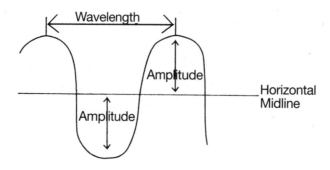

Figure 4

vertical line, rather than from the ceiling. Here, it is stated that this position is five degrees of arc (5°) from the vertical.

Another kind of concept diagram is called a wave diagram such as the one shown above (Figure 4). This diagram could represent sound waves heard in the air, electromagnetic waves from alternating currents, water waves, and so forth. Whenever we are talking about waves, we have to describe them by how strong they are or how high they are—this is their **amplitude**—and by the distance from one crest to the next crest, called the **wavelength**. The wavelength is not the distance between any two crests, or between a crest and a trough, but only between two crests one right after the other. The wave's amplitude is the distance from its midline to either a crest or trough. Water and sound waves do not carry along the substance through which they travel—at least, not very far at all. Electromagnetic waves in free space do not need any material to support them in their travel.

When electrical currents flow through wires and other devices like lamps and toasters, they are represented by diagrams that are called schematic diagrams. Two such schematic diagrams are shown below.

In the first diagram on the left (Figure 5), the jagged-line symbol represents any one of three resistances—R1, R2, R3—in the circuit. Scientists use a general symbol like R combined with numbers, called subscripts, to distinguish different resistances. In a similar way, the letter "V" is used to label the source of electrical "push," or voltage, from a battery. The symbol for a voltage source is a long-line segment and a parallel short-line segment nearby. These two straight pieces represent what are called the two poles of the battery. One of these poles, called the "negative" pole, is kept by the battery in such a state that there is always many more electrons built up on it than the other pole, called the "positive" pole. Therefore, electrons are pushed away from the negative pole of the battery, through the wire and the resistors, and toward the positive pole

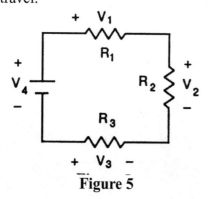

Figure 5

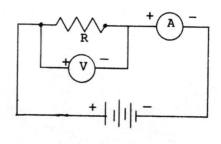

Figure 6

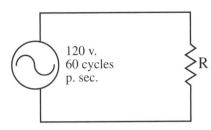

Figure 7

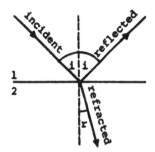

Figure 8

of the battery. This action completes the circuit through which current flows. The internal construction of the battery is such that it creates the electron imbalance necessary to keep the current flowing.

In Figure 6, the symbol of a circle with an "A" in it represents a device known as an **ammeter**. This electrical instrument is used to measure the amount of current, or **amperage**, that is flowing through the circuit. Notice that the two lines coming out from its sides are part of the in-line circuit wires. The ammeter is like a water flow meter and must be placed directly in the path of the current being measured. The circle with a "V" inside it represents another electrical instrument called a **voltmeter**. This device measures the amount of electrical push or **voltage** there is between any two points of the circuit. That is why it is shown with two lines coming out of it which are *not* in-line with the circuit itself. We want the voltmeter to touch the two points in the circuit we are interested in, but not to interfere with the current flow itself.

Notice that there are many alternating large and short lines in the second diagram where the voltage source exists. This is meant to indicate that the battery is a very strong one. The long and short lines between the end ones really don't matter except to indicate that the battery is strong. Only the two end segments, labeled as positive and negative, are important as connectors to the rest of the circuit. As always, solid straight lines represent the current-carrying wires of the circuit. Things that use electricity such as lamps, radios, toasters, and refrigerators are shown as general resistance R. This resistance may even be that of the wire itself, a property which all conductors, except those called superconductors, possess.

For example, consider the diagram above (Figure 7). The symbol ⊙ indicates an alternating voltage source. Even though the current may be alternating back and forth very rapidly in such a case, it will still produce heat in the resistances just as a direct current flow will. Next to the ⊙, the diagram usually states the peak voltage reached by the source in its variation and also the frequency at which it varies.

When talking about the direction of electromagnetic radiation, such as light, it is often useful to leave out the waves themselves and simply show a line representing the path of travel. A line representing direction for a wave is called a **ray**. A very concentrated beam of light is closely approximated by a ray in the real world. When we can study systems using just rays of light and not the waves themselves, scientists call it **geometrical optics**.

Suppose a ray of light hits a piece of material at an angle to its surface. The imaginary line through the surface and at 90° to it is called the **vertical** or **normal line** to the material face. See the diagram above (Figure 8).

The normal line is shown dashed in this diagram. Material #1 is air and material #2 is something denser—like water or glass. Notice that the incident ray splits into two parts at the surface. One part bounces off the surface and back into the first material (air). The second part continues on through the material, or medium, but is bent toward the normal line. The part that bounces back is called the **reflected ray** and the part that goes through is called the **refracted ray**. It always happens that the angle of incidence equals exactly the angle of reflection. This was first noticed by the Dutch physicist Willebrod Snell Van Royen in the early seventeenth century. It has since become known as Snell's Law. Notice that the diagram says nothing about the color of the waves being discussed, nor what the exact nature of the denser medium is, nor whether the medium is moving or still. The only thing assumed is that the two materials are the same consistency throughout.

Questions

Fill in the blanks.

1. If a diagram is drawn so that one unit of distance, like an inch, on the paper shows some other unit of distance in the real world, then we say that the diagram is drawn to _____.

2. _____ are used to represent items on a diagram.

Answers

1. scale

2. Symbols

DRAWING CONCLUSIONS

A scientific conclusion is a statement based on facts and logic regarding how something does happen or will happen. If the conclusion involves the future, then it is called a prediction.

For example, suppose a five-pound block rests on a flat table. It does not fall to the ground because the table supports it. Since the force of gravity would pull it down if the table wasn't there, we can logically conclude that the table must be exerting an upward force on the block to keep it at rest. This force must be upward because the force of gravity is downward. They must cancel each other in both magnitude and direction. This conclusion is not a prediction, but rather a statement about the condition of the block and table at any moment. But if we had a system such as the simple pendulum previously discussed, the bob could be held out at a certain short angle from the vertical and released. It would then be possible to draw a conclusion about when the bob and string would pass through the vertical from the moment of release. This conclusion is a prediction, and the drawing of predictive conclusions is one of the most important aspects of science as a human endeavor.

Drawing conclusions comes with much practice in looking at many aspects of a situation. As with reading a diagram, try to imagine the real-world conditions being spoken of and see how different things relate to each other. Look at the diagram on the next page (Figure 9) that represents the motion of our planet Earth around the sun.

Of course, this diagram is definitely not to scale! However, it does show several important ideas. One is that the orbit of the Earth around the sun is involved in the changing of the seasons. But it is not the distance from the sun that really causes these effects; rather, it is the $23\frac{1}{2}°$ tilt of the Earth's axis from the vertical that is most important to these changes.

The imaginary circular lines drawn over the Earth's surface are referred to as lines of latitude. Look at Position A. The imaginary line through the North and South Poles is called the spin axis of the planet. The Earth turns on this axis every 24 hours to produce one day and one night. The North Pole is the center point of the latitude known as the Arctic Circle. The latitude around the exact middle of the planet is called the equator. Sunlight coming straight at the Earth strikes the equator directly all day while the Earth is at or near Position A. The name associated with Position A is Equinox, which is Latin for "equal night." This means that there is an equal amount of day and night all over the world during this season.

Based on what you know, and looking again at the diagram, what conclusion can you draw about which other position of the Earth in its orbit results in the same conditions? The answer is the one shown as Position C. Here, the tilt is in the same orientation as Position A. At Position B, the North Pole points away from the sun and is in constant darkness, while the South Pole is in constant daylight. At Position D, this is just the opposite.

When people refer to the North polar regions as the "Land of the Midnight Sun," to what position in the diagram (Figure 9) do you think they are referring? Since the phrase "sun at midnight" really means that the sun is still shining in the sky when the hands of an evenly running watch read twelve o'clock midnight, we can logically conclude that this happens when the North Pole receives as much constant sunlight as possible. From the diagram, this happens only at the spot marked Position D. The two places in the Earth's orbit called the Solstices are where the times of longest days and longest nights occur.

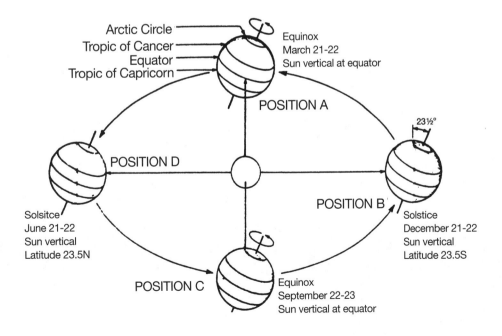

Figure 9

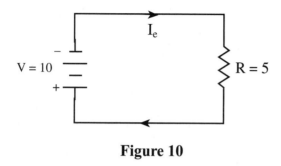

Figure 10

Question

1. During the Winter Solstice (Position B), what area on the Earth can you conclude becomes the "Land of the Midnight Sun?"

_____

_____

_____

Answer

1. The South Pole becomes the "Land of the Midnight Sun."

Physics allows conclusions to be drawn many times by means of a formula based on repeated observations in the past. In the case of an electrical circuit, the amount of electron current flowing is called (I_e), the voltage push is labeled V, and the resistance to current flow is simply labeled R. Using these symbols to represent possible numerical values, we can write Ohm's Law as the formula $I_e = V/R$. It is true for direct currents, but has a modified form for alternating currents.

Suppose we have the circuit (Figure 10) as shown above. The small "e" under the I shows that we are talking about the electron flow. Scientists say that the electrons have a "negative charge" and know that similar particles repel one another. This is what gives rise to the push of a battery in the first place — the

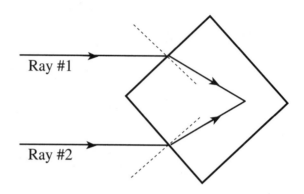

Figure 11

buildup of negative charges on the one terminal pushing away the electrons along the wire toward the opposite, or "positive," pole of the battery. The job of the device itself is to keep the two ends constantly charged positive and negative so that current can keep flowing. Using Ohm's Law, we could draw the conclusion that if there are 10 units of voltage and 5 units of resistance in a closed electrical circuit, then there are (10/5) = 2 units of amperage in current flowing through it.

We saw in the previous section that a light ray bends when it travels from one type of material into another. Going from a thinner to a thicker medium, the ray bends toward the vertical line at the interface. From this, we might logically conclude that light bends away from this imaginary perpendicular line when going from a thicker to a thinner material. This would be a logical conclusion and is, in fact, found to be true when tested in experiments.

To see some further consequences of this ability to draw conclusions based on formulas, consider the following situation: a glass cube in air is struck at the midpoint of two adjacent sides by parallel light beams as shown in the diagram (Figure 11). The two rays will bend at the surfaces toward the imaginary vertical line and, if the block is large enough, should meet at some point inside it. We can draw this conclusion, make it as a prediction, test it by experiment, and find that it is, in fact, correct.

So we see that there are several different types of conclusions possible in science and that they should all be drawn from facts and logic. This skill is probably the most important one in all our studies of people and things in the world around us.

Questions

Fill in the blanks.

1. A scientific _____ is a statement based on facts and logic regarding how something does happen or will happen.

2. Physics allows conclusions to be drawn many times by means of a(n) _____ based on repeated observations in the past.

3. Several types of conclusions are possible in science — they should all be drawn from _____ and _____.

Answers

1. conclusion

2. formula

3. facts, logic

TOPICS IN PHYSICS

This material will provide you with an overview of the physics field. These are topics you will need to become familiar with before taking the GED examination.

Motion

Moving objects can be measured for speed or momentum. Speed is the distance an object travels per unit of time. Cars measure speed in miles-per-hour (mph).

$$\text{Speed} = \frac{\text{distance}}{\text{time}}$$

If a car travels 3.5 km in 7 minutes, it has a speed of 3.5 km/7 min, or 0.5 km/min.

Momentum is the tendency of an object to continue in its direction of motion.

$$\text{Momentum} = \text{mass} \times \text{speed}$$

The heavier a moving object is, or the faster it is moving, the harder it is to stop the object or change its direction.

Energy and Work

Energy is the ability to do work. Energy comes in many different forms; for example, heat, light, and sound. All energy can be described as potential or kinetic. Potential energy is stored through chemical structure, position, or physical configuration. Kinetic energy is energy of motion. Light, sound, and heat are kinetic energy, as is the energy possessed by a moving object.

Energy can be transformed from one type to another, but it never is created or destroyed. The potential chemical energy in a peanut butter sandwich is transformed through digestion and metabolism into the kinetic energy of heat and motion. The potential energy of a book sitting on a shelf is turned into the kinetic energy of motion, sound, and heat as the book falls and hits the floor.

Heat is an important type of energy. Heat may travel through three paths: conduction, convection, and radiation. Conduction occurs when a hot material comes in contact with a cold one. Heat moves from a hot material into a cold material until the temperature of both is equal. An example of conduction is the heating of a metal spoon when it is used to stir a cup of hot tea.

Convection is based on a density change caused by heating. As materials, especially gases and liquids, are heated, they become less dense. Warm air, which is less dense, rises, while cold air, which is more dense, sinks. In a room or other enclosed space, this rising and falling of materials of different density creates a current of air (or other heated material). As heat is added to the space, from a source like a stove or sunny window, the current carries the heat through the space.

Radiation is heat that spreads out from a very hot source into the surrounding material. Radiant heat energy is carried by electromagnetic waves, just like the light given off by a hot light bulb filament. Radiant heat energy travels in straight lines in all directions from its source. Sources of radiant heat include wood stoves and light bulbs in homes.

Insulators are materials that slow down or prevent the movement of heat. Air is a good insulator. Most commercial insulation consists of a material with many pockets of air. Conductors are materials that transmit heat well. Metals are excellent heat conductors.

Work occurs when a force (push or pull) is applied to an object, resulting in movement.

$$\text{Work} = \text{force} \times \text{distance}$$

The greater the force applied, or the longer the distance traveled, the greater the work done. Work is measured in newton-meters or foot-pounds. One newton-meter equals one joule.

Mass is a measure of the amount of matter in an object. Weight is the gravitational force on an object. Mass is a constant; it never changes with location. Weight varies with the pull of gravity. Objects weigh less on the moon than on Earth. In space, where there is no gravity, objects are weightless (but they still have mass).

Power is work done per unit time.

$$\text{Power} = \frac{\text{work done}}{\text{time interval}}$$

If someone moves an object weighing 5 newtons over a distance of 10 meters in 30 seconds, they use the power of 1.7 watts.

$$\frac{5n \times 10m}{30 \text{ sec}} = 1.7 \text{ n-m/sec, or } 1.7 \text{ watts}$$

1 watt = 1 n-m/sec

Machines change the direction or strength of a force. Simple machines are used throughout our lives.

Simple Machine	Examples
inclined plane	ramp, wedge, chisel
screw	cork screws, jar lids
lever	seesaw, crowbar, jack
wheel and axle	doorknob, bicycle
pulley	fan belt, elevator

In designing machines, 100% efficiency is the goal.

$$\% \text{ efficiency} = \frac{\text{work done}}{\text{energy used}} \times 100.$$

100% efficiency can never be achieved, because some energy is always lost through friction or heat production.

Wave Phenomena

Sound and light are wave phenomena. Waves are characterized by wavelength, speed, and frequency. Wavelength is the distance between crests or troughs of waves. (See Figure 12 below)

Speed is how fast a wave crest or trough moves. If a wave moves 4 meters in 2 seconds, its speed is 2m/sec.

$$\text{speed} = \frac{\text{distance}}{\text{time}} = \frac{4m}{2 \text{ sec}} = 2\text{m/sec}.$$

Frequency is the number of crests or troughs that move past a point per second. Frequency is measured in Hertz. One wave moving past a point per second equals one Hertz.

$$\text{Frequency} = \frac{\text{speed}}{\text{wavelength}}$$

Sound is caused by the vibration of objects. This vibration creates waves of disturbance that can travel through air and most other materials. If these sound waves hit your eardrum, you perceive sound.

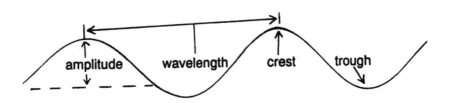

Figure 12

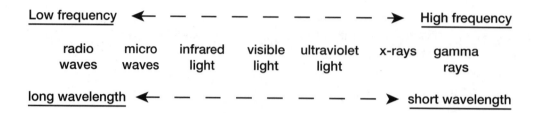

Figure 13

Sound is characterized by its pitch, loudness, intensity, and speed. Pitch is related to frequency. High pitches (e.g. high music notes) have high frequencies. Loudness is related to wave amplitude. Loud sounds have large amplitudes. Sound intensity is measured in decibels. Intensity is related to amplitude and frequency of sound waves. Loud, high-pitched music has a much greater intensity than quiet, low-pitched music.

The speed of sound waves is related to their medium. Sound travels more quickly through more dense materials (solids, liquids) than less dense materials (gases). Sound does not travel through a vacuum.

Light is a type of electromagnetic wave. The chief types of electromagnetic waves, listed according to their relative frequency and wavelength, are shown above. (Figure 13)

Light travels much more quickly (300,000 km/sec) than sound does (330 m/sec). It can pass through a vacuum. As light passes through a material, it travels in a straight path. When light moves from one material to another, it may be transmitted, absorbed, reflected, or refracted.

Transparent materials (e.g. water, glass) allow light to pass directly through them. This passing through is called transmission. Opaque objects (e.g. wood) absorb light. No light comes out of them. Mirrors reflect light. They re-emit light into the medium it came from. Light rays going into a mirror are called incident rays. Light rays going out of a mirror are called reflected rays. (See Figure 14)

Refraction is the bending of light. Light may be refracted when it moves from one material to another (e.g. air □→ water). Mirages are formed when light refracts while moving from cool air to warm air.

Sometimes, during refraction, sunlight is broken into the colors that form it, causing a spectrum. The colors in a spectrum are red,

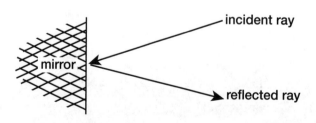

Figure 14

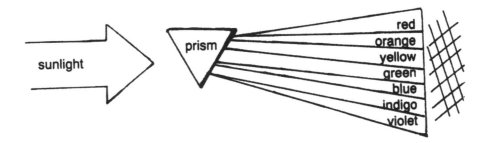

Figure 15

orange, yellow, green, blue, indigo, and violet. A rainbow is a spectrum caused when light passes from dry air into very humid air. (See Figure 15 above)

Lenses are transparent materials used to refract light. The shape of a lens determines how light passing through it will be bent. (See Figure 16 below)

Basic Electricity

All matter is made of atoms. All atoms contain positively charged particles, called protons, and negatively charged particles, called electrons. Protons are tightly bound to atoms and cannot move much. Electrons are loosely attached to atoms, and may leave one atom to join another.

Atoms may carry electrical charges. A neutral atom has equal numbers of protons and electrons in it. The charges of the protons and electrons cancel each other, so the atom has no net charge. If an atom has more electrons than protons, the extra electrons give the atom a negative charge. If an atom has less electrons than protons, the missing electrons leave the atom with a positive charge.

Electrons may not be destroyed. If two objects are rubbed together, however, electrons may move from one object to another, leaving both charged. Electrons may also flow through certain materials. The flow of electrons produces an electric current. Conductors are materials that let electrons flow freely (e.g. metals, water). Insulators are materials that do not let electrons flow freely (e.g. glass, rubber, air).

Electricity (electric current) flows from areas of many electrons to areas of few electrons. The path along which electrons flow is called a circuit. In a direct current (DC) circuit, electrons flow in one direction only. Alternating current (AC) is the type of current supplied over power lines. Alternating current changes direction many times per second.

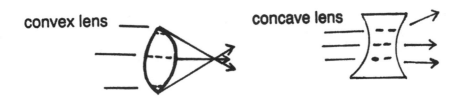

Figure 16

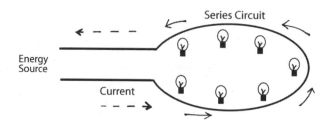

Figure 17: Series Circuit

Circuits may be described as being in series or in parallel. Series circuits are made of a single pathway, through which all current must flow. If any part of a series circuit breaks, the circuit is "opened," and flow of current must stop. Some sets of Christmas tree lights are designed in series. If one bulb in the string of lights burns out, none of the lights in the string will work, because current is disrupted for the entire string. (See Figure 17)

Parallel circuits provide more than one pathway for current to flow. If one of the pathways is opened, so that current cannot flow in it, the current will continue to move through the other paths. Most circuits, for example, those in our homes, are wired in parallel, so that burned out light bulbs and turned off television sets do not disrupt electricity used in other parts of our homes. (See Figure 18)

Fuses and circuit breakers are safety devices that limit the current flow in a circuit. Wires (lines) are limited in the amount of current they can safely carry. If too much current passes through them, they may heat up and melt or cause a fire. Current passing through lines increases with each appliance added to the circuit.

Fuses work by passing current through a thin metal ribbon. When the current exceeds the capacity of the fuse, the metal ribbon melts, leaving an open circuit, which cannot carry current. Circuit breakers use magnets and bi-metallic strips to open circuits if the current becomes too great. Fuses must be replaced after they "blow," or melt. Circuit breakers may simply be reset to be used again.

Volts measure the work done as electrons move from one point to another within a circuit. Battery "strength," or ability to do work, is measured by volts. Amperes measure the current, or flow of charge through a circuit. Ohms measure the resistance to the flow of electrons.

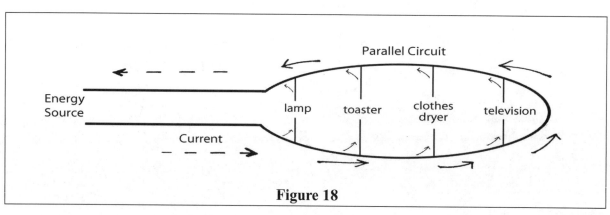

Figure 18

Volts = amperes × ohms

Watts measure electrical power consumption. Electrical appliances and light bulbs are rated by their wattages so consumers can compare power consumption before purchasing these products. One watt equals one joule per second of power (one newton-meter per second). One watt of energy can lift an object weighing one newton over one meter in one second. A kilowatt-hour is the amount of energy used in one hour by one kilowatt of power.

Power = current × voltage,

or 1 watt = 1 ampere × 1 volt

Magnetism

Magnets are solids that attract iron. Naturally occurring magnets are called lodestones. Magnetic forces make magnets attract or repel each other. Magnetic forces are created by regions in magnets called magnetic poles. All magnets have a north and a south pole. The north pole of one magnet will repel the north pole of another magnet; the same holds true for south poles. The south pole of one magnet will attract the north pole of another magnet.

A magnetic field is the area affected by magnetic force. A magnetic field surrounds both poles of a magnet. A magnetic field can be created by an electric current. Electromagnets create large magnetic fields with electric current. Similarly, if a wire is moved through a magnetic field, a current is produced. Electric generators make electricity by passing wires through a magnetic field.

The Earth has a magnetic field. Compasses are magnets that align themselves with Earth's magnetic field.

☞ Practice: Physics

DIRECTIONS: Fill in the blanks for questions 1-6. For questions 7-15, circle "T" if the statement is true, and circle "F" if the statement is false.

1. An action that results in something else happening is called a(n) _____.

2. The result of a cause is known as a(n) _____.

3. The momentum of an object depends on two things: _____ and _____.

4. The word describing how many electrons are in an electric current is _____.

5. Georg Ohm discovered the law relating to _____ electric currents.

6. The back-and-forth motion of electrons causes _____ to be emitted into free space.

7. Infer is the same as predict.

   T   F

8. All light travels at the same speed in free space.

   T   F

9. If an object has zero momentum, then it must be at rest.

   T   F

10. An effect always precedes its cause.

    T F

11. If a conclusion involves the future then it is called a prediction.

    T F

12. Drawing conclusions comes with much practice in looking at many aspects of a situation.

    T F

13. There is only one type of conclusion in science.

    T F

14. The flow of electrical current through wires and other devices can be shown on a diagram.

    T F

15. A scale key is needed to understand distance on a scaled diagram.

    T F

Answers

1. cause

2. effect

3. velocity, mass

4. amperage

5. direct

6. electromagnetic radiation

7. False. Infer means "to determine that something follows logically." Predict means "to state, tell about, or make known in advance on the basis of special knowledge."

8. True.

9. True.

10. False. Cause *must* come before effect.

11. True.

12. True. Drawing conclusions involves taking everything you have learned and coming up with an answer to a particular problem.

13. False. More than one conclusion can be drawn in science. For example, physics allows conclusions to be drawn many times by means of a formula based on repeated observations in the past.

14. True. There are many examples of diagrams that show the flow of electrical currents through wires and other devices.

15. True. Without a scale key, you could not understand distance on a scaled diagram.

REVIEW

In this chapter, you learned how to attack problems in physics by recognizing cause and effect, making inferences, reading diagrams, and drawing conclusions.

Recognizing cause and effect is similar to the skill of comparing and contrasting, which you learned in the chapter entitled "Chemistry," because you are evaluating the relationship between two things. This skill is different in that you are not evaluating all the similarities and differences between two objects, but rather seeing how they directly influence one another. This is an important concept in physics because you always want to know how one thing will affect another so you can expect certain outcomes.

Making inferences is a way of thinking outside the stated facts. You can take the facts you know and come up with other facts. In a way, what you are doing is combining what you have learned with what you already know.

Reading diagrams is important because it is the key to learning concepts in physics. Without diagrams it would be very hard to understand many of the concepts of physics, particularly how one thing can cause the action of another. Seeing things will always help you to better understand a physical concept.

Drawing conclusions involves taking everything you have learned and coming up with an answer to a particular problem. In a way you are acting like a detective, analyzing all the facts to produce the solution to a mystery. It is always necessary to really think through conclusions, since even well-trained scientists sometimes come up with the wrong conclusions. Many times you need to test our conclusions by applying them to other situations.

Science

Post-Test

SCIENCE

POST-TEST

DIRECTIONS: Carefully read and answer each of the following questions. Choose the best answer choice for each question.

1. The element which has the fewest numbers of parts is _____.
    (1) helium
    (2) carbon
    (3) hydrogen
    (4) nitrogen
    (5) gold

2. What type of information cannot be acquired by reading a map?
    (1) Weather forecasts
    (2) Earthquake damage probabilities for the next 100 years
    (3) The correlation between the burning of fossil fuels and global carbon dioxide levels
    (4) The types of rock formations within the United States
    (5) The location of a city

3. "To state, tell about, or make known in advance on the basis of special knowledge" is the definition of _____.
    (1) guess
    (2) predict
    (3) question
    (4) reason
    (5) comprehend

4. Substances made up of more than one element are called _____.
    (1) atoms
    (2) protons
    (3) electrons
    (4) neutrons
    (5) compounds

5. If you were to start out at an elevation of 1,200 feet on a map and you walked to an elevation of 1,750 feet, what amount of increase in elevation would you experience?
    (1) 950 feet
    (2) 500 feet
    (3) 450 feet

GED Science

(4) 650 feet

(5) 550 feet

6. In the context of the scientific method, an educated guess that fits as much of the known information as possible is called a _____.

   (1) belief
   (2) fact
   (3) opinion
   (4) statement
   (5) hypothesis

7. A substance produced by merely combining two or more elements without causing a chemical reaction and changing atomic bonds is called a(n) _____.

   (1) solid
   (2) mixture
   (3) liquid
   (4) aqueous solution
   (5) gas

8. What is released when the nucleus of an atom is split?

   (1) Oxygen
   (2) A liquid
   (3) Radiation
   (4) Odors
   (5) Water

9. A map that indicates longitude and latitude, as well as sub-measurement units such as minutes and seconds, would be useful for locating _____.

   (1) a city

(2) the Ring of Fire

(3) ocean currents

(4) an ocean

(5) a continent

10. A compound word is the _____ of two words into a(n) _____ word.

    (1) dissection, descriptive
    (2) combination, larger
    (3) combination, hyphenated
    (4) combination, complex
    (5) rephrasing, known

11. Any machine that includes two or more simple machines is called a(n) _____.

    (1) electronic machine
    (2) reproducible machine
    (3) solo machine
    (4) mechanical machine
    (5) compound machine

12. An example of a process that involves a chemical reaction is _____.

    (1) hammering a nail
    (2) splitting wood
    (3) throwing a ball
    (4) cooking fudge
    (5) mixing orange juice with water

13. The passing of traits from parents to their offspring is called _____.

    (1) nurture
    (2) chromosomes
    (3) heredity

(4) instinct

(5) intuition

14. In a sense, the whole sequence of events that occurs in the world is a(n) _____ chain.

    (1) unbreakable
    (2) finite
    (3) cause/effect
    (4) infinite
    (5) unknown

15. Finding details is important to the understanding of scientific concepts, theories, or presentations because _____ .

    (1) the key to understanding a scientific concept, theory, or presentation can be contained within the details
    (2) details will tell you the history of a scientific concept, theory, or presentation
    (3) details give a broad overview of a scientific concept, theory, or presentation
    (4) "details" is a word that means scientific understanding
    (5) scientific details are more important than details in other subjects.

16. Being able to _____ from different pieces of scientific information is most important to the understanding of a scientific concept, theory, or presentation.

    (1) formulate an opinion
    (2) make an observation
    (3) ask a question
    (4) draw a rational conclusion
    (5) make a summarization

17. A main theory of biology which states that all living things are composed of cells and all cells arise from other cells is called _____ .

    (1) The Theory of Relativity
    (2) The Theory of Evolution
    (3) Cosmic Theory
    (4) Cell Theory
    (5) Atomic Theory

18. What type of scientific information would be appropriate for presentation in a circle graph?

    (1) The life cycle of a hurricane
    (2) The number of tropical storms, hurricanes, and strong hurricanes that occurred in a given year
    (3) The location of a specific hurricane at a specific point in time
    (4) The strength of a specific hurricane
    (5) The amount of flooding caused by a specific storm

19. A distinction between plant and animal cells is that many of the cells of green plants contain _____.

    (1) proteins
    (2) chloroplasts
    (3) amino acids
    (4) mass
    (5) atoms

20. Chemical compounds that have only hydrogen and carbon with only single bonds are called _____.

    (1) amines
    (2) alkanes
    (3) chlorides
    (4) noble gases
    (5) salts

21. The parts of any diagram are connected with identifying words called _____.

    (1) labels
    (2) titles
    (3) captions
    (4) subscript
    (5) source

22. Burning occurs when a chemical element or compound reacts with _____.

    (1) chlorine
    (2) nitrogen
    (3) sodium
    (4) water
    (5) oxygen

23. The horizontal axis in a bar graph is usually called the _____ axis; the vertical axis in a bar graph is usually called the _____ axis.

    (1) Y, W
    (2) Y, X
    (3) J, Z
    (4) X, Y
    (5) Y, Z

24. All living things go through several stages of development called the _____.

    (1) adolescence
    (2) vital signs
    (3) biorhythms
    (4) life cycle
    (5) maturity level

25. Protons are _____ charged, and electrons are _____ charged.

    (1) negatively, positively
    (2) positively, not
    (3) positively, negatively
    (4) neutrally, positively
    (5) negatively, never

26. When living things and nonliving things interact within a certain area on the earth, scientists refer to that area as _____ _____.

    (1) a life zone
    (2) a residence
    (3) an ecosystem
    (4) a biosphere
    (5) a dwelling

27. Scientific opinions eventually are accepted as fact or are disproved by direct _____ or _____.

    (1) opinions, observation
    (2) experimentation, observation
    (3) ideas, statements
    (4) expressions, questions
    (5) intuitions, interpretations

28. The _____ of an object is the combination of its speed multiplied by how much matter is in it.

    (1) force
    (2) energy
    (3) momentum
    (4) friction
    (5) direction

29. It is possible to put all animals into categories known as _____.

    (1) stations
    (2) status
    (3) classes
    (4) levels
    (5) species

30. If the total amount of momentum before a collision equals the total amount of momentum after a collision, scientists say the collision was _____.

    (1) predictable
    (2) inelastic
    (3) conservative
    (4) elastic
    (5) severe

31. _____ is the flow of particles called electrons.

    (1) Sound
    (2) Electricity
    (3) Light
    (4) Electrolysis
    (5) Electrons

32. The strength or height of a wave is called its _____.

    (1) magnitude
    (2) latitude
    (3) longitude
    (4) amplitude
    (5) heading

33. The distance from one wave crest to another is called the _____ of the wave.

    (1) peak length
    (2) wavelength
    (3) pitch
    (4) level
    (5) crest

GED Science

SCIENCE

ANSWER KEY

1. (3)	10. (2)	19. (2)	28. (3)
2. (3)	11. (5)	20. (2)	29. (5)
3. (2)	12. (4)	21. (1)	30. (4)
4. (5)	13. (3)	22. (5)	31. (2)
5. (5)	14. (3)	23. (4)	32. (4)
6. (5)	15. (1)	24. (4)	33. (2)
7. (2)	16. (4)	25. (3)	
8. (3)	17. (4)	26. (3)	
9. (1)	18. (2)	27. (2)	

POST-TEST SELF-EVALUATION

Question Number	Subject Matter Tested	Section to Study (section, heading)
1.	main idea	III, Finding the Main Idea
2.	maps	IV, Reading a Map
3.	predictions	II, Making Predictions
4.	main idea	III, Finding the Main Idea
5.	maps	IV, Reading a Map
6.	scientific method	II, Making Predictions
7.	main idea	III, Finding the Main Idea
8.	main idea	III, Finding the Main Idea
9.	maps	IV, Reading a Map
10.	compound words	IV, Understanding Compound Words
11.	cause-and-effect relationships	V, Recognizing Cause and Effect
12.	comparisons	III, Comparing and Contrasting
13.	predictions	II, Making Predictions
14.	cause-and-effect relationships	V, Recognizing Cause and Effect
15.	details	IV, Finding Details
16.	conclusions	IV, Drawing Conclusions
17.	inferences	II, Making Inferences
18.	circle graphs	IV, Reading a Circle Graph
19.	inferences	II, Making Inferences
20.	chemical compounds	III, Roots, Suffixes, and Prefixes
21.	diagrams	II, Understanding a Diagram
22.	predictions	III, Making Predictions
23.	bar graphs	IV, Reading a Bar Graph
24.	information	II, Summarizing Information
25.	chemical formulas	III, Topics in Chemistry
26.	cause-and-effect relationships	II, Recognizing Cause and Effect
27.	facts and opinions	II, Fact vs. Opinion
28.	inferences	V, Making Inferences

II = Life Sciences   III = Chemistry   IV = Earth Science   V = Physics

GED Science

Question Number	Subject Matter Tested	Section to Study (section, heading)
29.	categories	II, Understanding a Photograph
30.	inferences	V, Making Inferences
31.	inferences	V, Making Inferences
32.	diagrams	V, Reading a Diagram
33.	diagrams	V, Reading a Diagram

II = Life Sciences III = Chemistry IV = Earth Science V = Physics

POST-TEST ANSWERS AND EXPLANATIONS

1. **(3)** Hydrogen (3) has only two parts, a proton nucleus surrounded by one electron. Helium (1), carbon (2), nitrogen (4) and gold (5) occur after hydrogen on the periodic table of the elements and have more protons, neutrons, and electrons than hydrogen.

2. **(3)** Weather forecasts (1), earthquake damage probabilities for the next 100 years (2), the types of rock formations within the United States (4), and the location of a city (5) can be obtained by reading a map. The correlation between the burning of fossil fuels and global carbon dioxide levels (3) cannot be presented on a map. A graph would be a more appropriate means of presenting this information.

3. **(2)** "To state, tell about, or make known in advance on the basis of special knowledge" is the definition of predict (2). To make a guess (1) is not based on knowledge, but rather on a person's intuition or feeling. To question (3) refers to the desire to gain knowledge. To reason (4) is to think something through rationally. To comprehend (5) is to understand.

4. **(5)** Compounds (5) are made up of more than one element. Atoms (1) are the basic building blocks of chemicals. Protons (2), electrons (3), and neutrons (4) are constituents of atoms.

5. **(5)** To determine the answer, subtract your original elevation, 1,200 feet, from your final elevation, 1,750 feet. The result is 550 feet (5). Choices (1), (2), (3), and (4) are incorrect.

6. **(5)** In the context of the scientific method, an educated guess that fits as much of the known information as possible is called a hypothesis (5). A belief (1) is a personal idea or conviction based upon a person's observations, value system, or faith. A fact (2) is something that is considered a provable and widely accepted idea. An opinion (3) is based upon a personal belief or idea. A statement (4) is a point of view that is expressed either orally or is written.

7. **(2)** A mixture (2) is a substance made by combining more than one element without causing a chemical reaction and changing atomic bonds. Solids (1), liquids (3), and gases (5) can either be comprised of a pure element or can be a mixture of elements. An aqueous solution (4) is a solution in which water is a major constituent.

8. **(3)** Radiation (3) is released when the nucleus of an atom is split. Oxygen (1), a liquid (2), odors (4), and water (5) are not by-products of the splitting of the nucleus of an atom.

9. **(1)** A map with precise longitude and latitude units would be ideal for locating something that occupies a distinct point on a map like a city (1). The Ring of Fire (2), ocean currents (3), an ocean (4), and a continent (5) cover huge areas on a map and are not easily indicated in precise units of longitude and latitude.

10. **(2)** A compound word is the combination of two words into a larger word (2). Dissection, descriptive (1), combination, hyphenated (3), combination, complex (4), and rephrasing, known (5) do not properly define a compound word.

11. **(5)** Any machine that includes two or more simple machines is called a compound machine (5). An electronic machine (1), a reproducible machine (2), and a mechanical machine (4) can be either a simple machine *or* a compound machine. A solo machine (3) is an independent, individual machine.

12. **(4)** Cooking fudge (4) involves the reaction of many of the ingredients with oxygen. Hammering a nail (1), splitting wood (2), and throwing a ball (3) are physical activities that do not involve chemical reactions. Mixing orange juice with water (5) results in a mixture which does not involve a chemical reaction.

13. **(3)** The passing of traits from parents to their offspring is called heredity (3). Chromosomes (2) are rod-shaped bodies in cells which relate to heredity. Nurture (1) refers to shaping a person by influences from the outside world. Instinct (4) is a phenomena in which a species intuitively knows a survival skill. Intuition (5) is an inner knowledge or "gut feeling" which guides a living being to make a decision or take an action.

14. **(3)** In a sense, the whole sequence of events that occurs in the world is a cause/effect (3) chain. Unbreakable (1) refers to a chain that cannot be broken. A finite (2) chain refers to a chain that has a definite ending. An infinite (4) chain refers to a chain that does not have a definite ending. Unknown (5) chain means that it is not known; however, there are many events that occur in the world which are known.

15. **(1)** The key to understanding (1) can be found in details. Details do not always present the history (2) of scientific concepts, theories, and presentations. A broad overview (3) is not detailed, but general. The word "details" does not mean scientific understanding (4). Scientific details are not necessarily more important than details in other subjects (5).

16. **(4)** Formulating an opinion (1) about scientific information does not always enhance the understanding of a scientific concept, theory, or presentation. Making an observation (2) can be useful, but is not of paramount importance to the understanding of a scientific concept, theory, or presentation. Asking a question based upon pieces of scientific information (3) is important, but is not most important to the understanding of a scientific concept, theory, or presentation. Being able to draw a rational conclusion (4) is of paramount importance to the understanding of scientific concepts, theories, or presentations. To make a summarization (5) of scientific information restates information and does not always demonstrate a clear understanding of scientific concepts, theories, or presentations.

17. **(4)** A main theory of biology that states that all living things are composed of cells and all cells arise from other cells is called Cell Theory (4). The Theory of Relativity (1) is a theory of mass, space, and time. The Theory of Evolution (2) is a theory which explains how life forms on Earth developed. Cosmic Theory (3) is a generic term for theories which attempt to explain the workings of the universe. Atomic Theory (5) explains the workings and laws governing atoms and their associated protons, neutrons, and electrons.

18. **(2)** The life cycle of a hurricane (1) cannot be presented in a circle graph. It would be best presented in written form with diagrams. The location of a specific hurricane at a specific point in time (3) cannot be presented in a circle graph. It would be best presented on a longitude and latitude map. The strength of a specific hurricane (4) cannot be presented in a circle graph. The amount of flooding caused

by a specific storm (5) cannot be presented by a circle graph. The number of tropical storms, hurricanes, and strong hurricanes that occurred in a given year (2) could easily be presented in a circle graph.

19. **(2)** A distinction between plant and animal cells is that many of the cells of green plants contain chloroplasts (2). Proteins (1) are constructed of amino acids (3). Proteins and amino acids can be found throughout a living plant or animal. Mass (4) is the amount of matter contained in a material. All cells are composed of atoms (5).

20. **(2)** Chemical compounds that have only hydrogen and carbon with only single bonds are called alkanes (2). Amines (1), chlorides (3), noble gases (4), and salts (5) are distinct groups of chemicals not related to alkanes.

21. **(1)** The parts of any diagram are connected with identifying words called labels (1). Titles (2) summarize the information found in a diagram. Captions (3) provide a brief explanation about the interpretation of a diagram. Subscript (4) is a brief piece of information at the bottom of a diagram which provides details concerning the diagram or a portion of the diagram. The source (5) of a diagram is often provided at the bottom of the diagram as a subscript.

22. **(5)** Burning occurs when a chemical element or compound reacts with oxygen (5). Although chlorine (1), nitrogen (2), sodium (3), and water (4) can react with other elements or compounds, their reactions are not referred to as burning.

23. **(4)** Y is not the horizontal axis, W is not an axis (1). Y is not the horizontal axis, X is not the vertical axis (2). J and Z are not axes (3). Y is not the horizontal axis, Z is not an axis (5). X is the horizontal axis, Y is the vertical axis (4).

24. **(4)** All living things go through several stages of development called the life cycle (4). Adolescence (1) is a stage of development between childhood and adulthood. Vital signs (2) are the biological indicators that an animal is alive, such as pulse and blood pressure. Biorhythms (3) are a number of interrelated biological and psychological indicators, such as mood and energy level. Maturity level (5) is an abstract measure of a human's attention span, rationality, and manners.

25. **(3)** Protons are positively charged; electrons are negatively charged (3). Choices (1), (2), (4), and (5) do not correctly indicate the charge of protons and electrons.

26. **(3)** When living things and nonliving things interact within a certain area on the Earth, scientists refer to that area as an ecosystem (3). A life zone (1) is a generic term that could apply to any area with living things. A residence (2) is a location which a living thing occupies for shelter. The biosphere (4) is a general term for the area in the Earth's atmosphere that supports life. A dwelling (5) is a location in which a particular animal or human finds shelter.

27. **(2)** Scientific opinions are eventually accepted as fact or are disproved by direct experimentation or observation (2). Opinions (1) cannot be utilized to discredit other scientific opinions. Ideas or statements (3) could be formulated to disprove a scientific opinion, but would have to be supported with experiments or observations. Expressions and questions (4) can be stated or raised concerning scientific opinions, but cannot be solely utilized to accept or disprove scientific opinions. Intuitions and interpretations (5) are unique to

individuals and cannot be used to accept or disprove scientific opinions.

28. **(3)** The momentum (3) of an object is the combination of its speed multiplied by how much matter is in it. Force (1) is the amount of pressure needed to change an object's speed. Energy (2) is the ability of an object to do work. Friction (4) causes an object to slow down due to contact with another object. Direction (5) is the place toward which an object is moving.

29. **(5)** It is possible to put all animals into categories known as species (5). Stations (1), status (2), classes (3), and levels (4) refer to a tendency to rank things, including animals, according to their perceived value or worth.

30. **(4)** If the total amount of momentum before a collision equals the total amount of momentum after a collision, scientists say the collision was elastic (4). An inelastic (2) collision is the opposite of an elastic collision. A conservative (3) collision is a generic term for a small collision. A predictable (1) collision is one in which the result of the collision can be calculated in advance. Severe (5) is a descriptive word which indicates that a collision was very strong.

31. **(2)** Electricity (2) is the flow of particles called electrons. Sound (1) consists of waves that are received by the ear. Light (3) consists of waves that are received by the eye. Electrolysis (4) is the splitting of water into hydrogen and oxygen utilizing electricity. Electrons (5) are negatively charged atomic particles.

32. **(4)** The strength or height of a wave is called its amplitude (4). Latitude (2) and longitude (3) are units of measurement used in geography. Magnitude (1) is another term for the strength of a wave; however, it does not indicate height. Heading (5) is the direction in which something is traveling.

33. **(2)** The distance from one wave crest to another is called the wavelength (2) of the wave. The peak length (1) means the greatest length of a wave. The pitch (3) refers to the level at which a sound wave is heard. The level (4) of a wave refers to the intensity or loudness of a sound wave. The crest (5) of a wave is the highest point or peak of a wave.

Science

Appendix: Glossary of Terms

SCIENCE

APPENDIX: GLOSSARY OF TERMS

A.K.A.—also known as.

Acid—a substance which contains a high concentration of hydrogen ions (pH less than 7); it is sour to the taste and reacts with bases (alkalis) to form salt and water.

Adaptation—the gradual process of adjustment to new physical conditions exhibited by living organisms.

Adjacent—lying close to; next to.

Adult—an organism that has reached full growth and full development.

Adult Moth—a mature moth that is able to fly.

Aeon—an infinitely long period of time.

Affected—acted upon; produced a change in.

AIDS—(Acquired Immune Deficiency Syndrome) a severe immunological disorder caused by the retrovirus HIV, resulting in cell-meditated immune response that is manifested by increased susceptibility to opportunistic infections and rare cancer.

Algae—chlorophyll-containing, mainly aquatic eukaryotic organisms, ranging from microscopic to multicellular, that lack true roots, stems, and leaves.

Alkane—a hydrocarbon having a single bond between the carbon and hydrogen atoms.

Allergic—having an abnormal reaction to a previously encountered allergen.

Allergy—hypersensitivity to a reintroduced allergen.

Almanac—calendar of days, weeks, and months to which astronomical data are added.

Alternating Current (A.C.)—electrical current that reverses direction at constant intervals; in the U.S.A., 60 times per second.

Amino Acids—class of organic compounds containing amino and carboxyl groups, basic constituents of proteins.

Ammeter—device that measures the rate of flow of an electrical current.

Amnion—the inner membrane of the amniotic cavity of the womb.

Amniotic Cavity—the separate enclosure inside the womb where nourishment of the embryo and fetus takes place.

Ampere—unit of electrical current; flow of electrons.

Amplitude—width, extent, or size.

Animal Pollinator—refers specifically to birds or flying insects that carry pollen to help in plant reproduction.

Antagonistic Muscle—a muscle that opposes skeletal movement.

Anther—the pollen-bearing organ on the tip of the stamen of a flower.

Anthracite Coal—a hard form of coal that burns with very little smoke or flame.

Antibody—protein in blood which counteracts the growth and harmful actions of bacteria and other foreign matter.

Appendicular Skeleton—all of the bones in the peripheral body; the skeleton of the limbs of the body

Arctic Circle—the point of latitude that the North Pole centers upon.

Artery—thick-walled muscular vessel which transports blood from the heart to the body.

Artificial Selection—the creation of new breeds of plants or animals due to human intervention (breeding).

Assumption—the act of taking something as true without proof.

Atmosphere—the mass of air, clouds, gases, and vapors surrounding the earth.

Atom—the smallest component of an element having the chemical properties of that element; they are the building blocks of matter.

Axial Skeleton—the bones that occur along the axis of the spine, including the skull.

Axle—a bar of wood or metal on which a wheel or system of wheels turn.

Bacteria—any of the unicellular, prokaryotic microorganisms of the class Schizomycetes which vary in terms of morphology, oxygen, nutritional requirements, and motility. They may be free-living, saprophytic, and pathogenic (the latter causing disease in plants or animals).

Bale—a package, compactly compressed in a protecting cover.

Ball-and-Socket Joint—a joint where the rounded end of a bone fits into a similarly shaped socket of another bone; e.g., the hip.

Bar Graph—a visual drawing in which rectangles are used to portray data.

Base—an alkaline substance having a pH greater than 7; forms a salt and water when combined with an acid.

Battery—a device that stores electrical energy by converting chemical energy into electrical energy.

Biology—the study of life; the comprehensive term for the science of living organisms and life processes which includes structure, function, and behavior.

Bud—the short sprout on a plant containing an underdeveloped leaf, branch, or flower.

Burn—to undergo combustion, the result of a chemical compound that combines with oxygen.

Butane (C_4H_{10})—a natural gas, found dissolved in crude oil, used as a fuel.

Cancer—malignant growth or tumor.

Capricious—fickle, illogical change of feelings or opinions; unpredictable.

Caption—a brief title or explanation to an article or chapter.

Carbon Dioxide (CO_2)—colorless, odorless gas formed during respiration, combustion, and organic decomposition.

Cardiac—concerning the heart.

Cardiac Muscle—muscle tissue found only in the heart.

Carnivore—a flesh-eating animal.

Carpel—a simple pistil; the seed-bearing part of a plant.

Cartilage—a firm, elastic, and flexible connective tissue of a translucent whitish color.

Category—a broad division or classification.

Caterpillar—the larva of butterflies and moths.

Cause-and-Effect—a relationship between action and result.

Cell—the microscopic and basic structural unit of all organisms.

Cell Theory—the theory that states: 1) All living things are composed of cells, and 2) All cells arise from other cells.

Cellulose—the chief constituent of the cell walls in plants.

Celsius Scale—method of temperature measurement with the freezing point of water fixed at 0° and the boiling point fixed at 100°.

Charles' Law—the volume of a gas, at constant pressure, will vary with a change in temperature ($V1/T1=V2/T2$).

Chemistry—the science that deals with the composition and properties of substances.

Chlorophyll—green pigment found in plant leaves that is essential to the photosynthetic process.

Chloroplasts—structure in plant cells that contain chlorophyll.

Cholesterol—a sterol found in all animal tissue; also in bile, gallstones, blood, and the brain; functions chiefly as a protective agent for skin.

Cholesterol Plaque—a formation, usually within arteries, due to an excessive amount of cholesterol in the blood that prevents normal flow.

Chorion—the outermost fetal membrane.

Chromosomes—tiny thread-like structures in the nucleus of a cell that contain information about and functions in the transmission of the cell's characteristics (and to a greater degree, the organism); the normal number of chromosomes in a human is 46.

Circle Graph—compares statistics by arranging the information in a circle and representing each amount by cutting out pieces of the circle, similar to the way a pie is cut.

Circuit—the path of an electrical current.

Cite—to quote or bring as proof.

Closed System—a release of energy plus heat that is contained and can be measured.

Cluster—a number of things growing or arranged together.

Coccyx—four incompletely formed vertebrae which form the lower extremity of the spinal column.

Collision—an interaction between particles in which momentum is conserved.

Combustion—a chemical combination accompanied by the production of heat and light.

Compare—to examine in order to note similarities and differences.

Compare and Contrast—showing the relationship of how selected items are similar and how these same items have different qualities.

Compound—a substance made up of two or more elements held together by chemical bonds.

Compound Machine—a machine that has two or more moving parts.

Concave—curving inward, rounded; opposite of convex.

Concept—a mental impression of an object or idea.

Conclusion—the result of an experiment, when opinions or decisions are reached.

Conduction—the passage of energy, sound, or electricity through a medium (the conductor) while the medium itself does not move.

Conservation of Energy—the total amount of energy fed into a system must equal the amount of energy released from the system plus any heat created.

Contagious—a disease or condition that can be transmitted by direct or indirect contact.

Context—word or words surrounding an unfamiliar word that provide clues to each word's meaning.

Contour—a line that shows elevation on a topographical map.

Contraction—movement bringing together.

Contrast—selecting items to find a difference.

Convection—any transportation of energy or matter by currents of matter; based on a density change caused by heating.

Convex—curving outward; opposite of concave.

Coordinates—a set of two or more magnitudes used to determine the position of a point, line, or plane.

Corrode—to wear away gradually (especially by chemical action).

Corrosive—able to corrode.

Cycle—a repeating series of events.

Cylinder—a roller-like body having straight sides with equal parallel circles at the ends.

Cytoplasm—the liquid-like cell substance between the cell membrane and the nucleus which contains the organelles of the cell.

Darwin, Charles—scientist associated with the theory of evolution and natural selection.

Data—specific facts, statistics, or information.

Dead Sea—a body of water in Israel that contains an overabundance of salt and is rich in minerals; only algae can live in it.

Deduce—to reach a conclusion by making assumptions.

Deoxygenated Blood—blood that is oxygen-deficient; carried by the pulmonary artery to the lungs.

Depict—to characterize or describe in words.

Deplete—to empty; to diminish.

Depression—1) Geology: a sinking or falling in of a surface. 2) Meteorology: an area of low barometric pressure.

Diagram—a drawing or figure used to explain how something works and what it looks like.

Direct Current—a current that moves in one direction.

Diverse—heterogenous; different kinds; several.

Dominant—ruling or prevailing; in heredity, a dominant trait is that which is always inherited.

Down's Syndrome—a form of mental retardation; another name is Mongolism or Trisomy 21.

Dwarf—an animal or plant abnormally small in size.

Earthquake—a sudden, violent shaking of the Earth's surface caused by the splitting or sliding of an underground rock formation.

Ecosystem—the interaction of living things and non-living things within a certain area on the earth.

Egg—a plant or animal cell containing the hereditary material of the female parent.

Elastic—capable of returning to an original form after deformation.

Electromagnetic Radiation—a magnetic field caused by the flow of alternating current.

Electromagnetism—the phenomenon associated with electric and magnetic fields (from the wiggle of the compass needle to the use of X-rays in a physician's office).

Electron—negatively charged particle that revolves around an atom's nucleus; has the smallest mass of any particle.

Element—a substance which cannot be separated into simpler substances by chemical means.

Embryo—the young of an animal in the early stages of development in the womb (in humans, from two weeks after fertilization to the end of the seventh or eight week)

Endangered Species—plants or animals on the verge of extinction.

Energy—the ability to do work.

Environmental—pertaining to the surroundings of a habitat.

Epicenter—the place where an earthquake originates.

Equator—the imaginary line around the Earth, equally distant from the North and South poles.

Equinox—the time at which the sun's rays cross the equator; when there is an equal amount of daylight and darkness in one day.

Evolution—the scientific theory stating that higher forms of life developed from simple and basic forms.

Experiment—a test made to demonstrate a principle or to discover something unknown.

Extinct—no longer in existence.

Fact—a statement that is always true; something that has been verified.

Factor—any one of the causes of a given result.

Fahrenheit Scale—method of temperature measurement with the freezing point of water fixed at 32° and the boiling point fixed at 212°.

Fault—a break or flaw in underground layers of rock.

Fecal—pertaining to feces (solid body waste).

Fertilization—act of insemination, impregnation, or pollination.

Fetus—the young of vertebrate animals between the embryonic and independent stages (birth).

Force—an influence on a person or object that produces or tends to produce a change in movement or in shape.

Filament—the stalk of a stamen that supports the anther.

Fission—in nuclear physics, the splitting of an atom into two fragments accompanied by a great energy release.

Flower—the part of a seeded plant that contains reproductive organs.

Forefront—the foremost place; primary place.

Formula—the series of symbols denoting the component parts of a substance.

Fossil—trace of a prehistoric animal or plant preserved in the earth, in rocks, or in caves.

Fossil Fuel—nonrenewable energy (basically, crude oil) formed from the fossils of living organisms that existed over 1 million years ago.

Frequency—the number of vibrations per unit time (usually 1 second) of a recurring phenomenon.

Friction—the surface resistance to relative motion.

Front—(weather) the border between two large air masses that are at different temperatures.

Fulcrum—the pivot point of a lever.

Gaseous—in the form of gas.

Gear—a toothed wheel that works to transmit power or change timing.

Gear Drive—a system in which one gear turns the other.

Gene—the hereditary factor that occupies a fixed location on a chromosome and is given by each parent to his/her offspring that determines hereditary characteristics.

Genetic Screening—medical tests given to a person to determine the presence of inherited diseases.

Genetics—scientific study of the hereditary of individuals.

Geologic—having to do with the science of the earth's crust, rocks, etc.

Geometrical Optics—applies to the study of the effects of lenses and mirrors on light beams.

Germ—a microorganism, can be disease-producing.

Germinate—to begin to grow or develop.

Global Warming—the increase in the average temperature on earth due to the accumulation of carbon dioxide (CO_2) in the upper atmosphere.

Glossy—something that is smooth and shiny.

Graph—a diagram showing the relationship between two or more quantities.

Gravity—the force of attraction of one body to another.

Greenhouse Effect—the accumulation of CO_2 in the upper atmosphere that acts to trap heat on the earth's surface (see Global Warming).

Greenwich Time—the basis for calculating standard time everywhere.

Heart Attack—a serious condition causing damage to an area of the heart due to a deprivation of oxygen.

Helium—an inert, nonflammable, and elemental gas.

Heredity—characteristics, traits, and qualities that are transmitted from parents to offspring.

Heterozygote—an individual with two different genes for a single trait (Bb).

Homozygote—an individual with the same genes for a single trait (bb).

Host—an organism that provides nourishment for a parasite.

Huntington's Disease—the result of a genetic flaw characterized by dementia, jerky movements, and ultimately death.

Hurricane—a violent storm with rain and winds higher than 70 miles per hour that frequently originates over warm tropical seas and moves over the North Atlantic Ocean, the Caribbean Sea, the Gulf of Mexico, the eastern North Pacific Ocean, or the western South Pacific Ocean.

Hybrid—the offspring of two animals or plants of different species (e.g. a mule is bred from a horse and a donkey).

Hypothesis—a tentative theory used to explain facts and to guide in the investigation of other facts.

Identify—to prove to be; to recognize as.

Imply—to suggest.

Industrial Revolution—the changes in economic and social organization (1760 England), characterized by the replacement of hand tools with power-driven machines and the concentration on mass production methods.

Inelastic Collision—a collision between two objects that results in a release of heat.

Infection—a contamination of the body by microorganisms under conditions which allow injury and disease to tissue.

Infectious—capable of causing infection.

Inference—to come to a conclusion.

Inheritance—the acquiring of characteristics or qualities of parents by their offspring.

Inorganic—compounds which do not contain carbon.

Interbreed—to breed by crossing different kinds of animals or plants (produces hybrids).

Interface—border between two objects.

International Date Line—line which separates one day from the next.

Internode—the area of a plant stem between the area where leaves are attached.

Investigation—a careful examination.

Involuntary Muscle—muscles that cannot be consciously controlled (e.g. the heart).

Iridium—a rare metallic element found deep within the earth's core.

Joint—the point where two or more bones come together.

Joule—the basic unit of energy in the MKS system.

Key—the explanation of abbreviations and symbols on a map.

Kinetic Energy—the energy contained in a moving object.

Label—words that identify and describe parts of a diagram.

Laboratory—place used for experiments or research in science.

Lamprey—an eel-like fish having a circular mouth and sharp teeth used to bore into the flesh of other fish and feed on their blood.

Landfall—sighting of land by a ship at sea; where a hurricane moves from the ocean to land.

Larva—an insect in the wingless, often worm-like (preadult) stage.

Latitude—the relative distance north or south of the equator; measured in degrees.

Lemur—any of various small, nocturnal tree-dwelling mammals found in Southern Africa and Madagascar, having large eyes and a fox-like face.

Lever—a simple machine, consisting of a bar pivoted at one point (fulcrum), used to move heavy objects.

Lichen—an organism, composed of a fungus in symbiotic union with an algae.

Life Cycle—the development stages of plants and animals.

Line graph—a visual picture that shows the relationship between two quantities.

Logic—the science of reasoning.

Longitude—the relative distance east or west on the Earth's surface measured in degrees; the zero point is fixed on a line called the Prime Meridian, which runs north and south through Greenwich, England.

Lysis—the dissolution of cells (destruction) by lysins.

Machine—something that helps promote work.

Magma—a complex mixture of solid and liquid rock and gases deep within the Earth; magma becomes lava after a volcanic eruption.

Magnet—a substance composed of iron or steel which has had its domains aligned to produce a magnetic field.

Manhattan Project—a secret project of the U.S. government, during World War II, to develop an atomic bomb.

Map Key—symbols that allow you to find certain information about places shown on a map.

Mass—a measure of the amount of matter in an object.

Matter—material capable of occupying space.

Mechanism—the structure of a machine or piece of machinery.

Medium—in bacteriology, a substance used for the cultivation of bacteria.

Membrane—a thin layer of tissue covering or dividing an organ; the boundary of a cell.

Mendel, Gregor—scientist responsible for hypothesizing that hereditary traits occur in pairs.

Mesh—to become interlocked, as gears of a machine.

Meteorite—a body of rock or metal which has reached the Earth from outer space (but has not yet fallen into the atmosphere).

Methane (CH_4)—a flammable, hydrocarbon gas.

Microorganism—a microscopic organism.

Midnight Sun—the term given when there is a period of sunlight visible at night in arctic and antarctic regions.

Miracle Drug—wonder drug that goes beyond the laws of nature to save lives.

Mixture—a combination of chemicals or substances that are not chemically bound to each other.

Molecule—a tiny mass of matter made up of two or more atoms.

Molten—liquefied metal or rock.

Momentum—the impetus of a moving body; Momentum = Mass × Velocity.

Moth—a nocturnal winged insect.

Muscle—a tissue composed of cells or fibers, the contraction of which produces movement in the body.

NaOH—an alkaline chemical, sodium hydroxide.

Natural Gas—colorless, flammable product (mostly methane) occurring in association with mineral oil deposits.

Natural Resource—any material found on Earth that exists independent of human industry (e.g., water, fish, wild game) and is somehow consumed by humans.

Navel—the depression in the center of the abdomen where the umbilical cord from the mother's placenta had been attached.

Negative—indicating opposition, resistance, or movement in a reverse direction; a quantity less than zero.

Neutron—one of the minute particles within the nucleus of an atom; has no electrical charge.

Niche—the area within a habitat occupied by an organism.

Node—a slightly enlarged portion of a plant stem where leaves and buds arise, and where branches originate.

North Pole—the northernmost point on Earth, located at 90° North latitude.

Nuclear Chemistry—the study of the properties of substances at the atomic level.

Nucleus—the core of an atom, containing protons and neutrons.

Nurture—to support and foster during the period of development.

Nutrient—anything that nourishes and promotes growth.

Nymph—the young of an insect that undergoes incomplete metamorphosis.

Observation—the act of watching, examining, and noting.

Ohm's Law—the voltage in an electrical circuit equals the current times the total resistance ($V=IR$).

Open System—a release of energy where heat escapes and cannot be measured.

Opinion—a belief that rests on grounds insufficient to produce complete certainty.

Orbit—the course followed by one heavenly body around another (e.g., the Earth's path around the sun).

Organic—pertaining to a class of chemical compounds that formerly comprised only those existing in plants or animals, but now includes all compounds of carbon.

Ovary—the female reproductive gland in which the ova and hormones that regulate female secondary sex characteristics develop.

Ovule—the structure in a plant or animal that contains an egg cell which develops into a seed or egg.

Oxygen (O_2)—a colorless, odorless gas found in air that supports animal life and combustion.

Oxygenated—combined or treated with oxygen.

Oxygenated Blood—blood that contains oxygen.

Ozone (O_3)—a condensed and very active form of oxygen containing three atoms; it has a very peculiar, pungent odor.

Parallel—lines or objects that are equally distant from each other at all points.

Parasite—an organism that survives by living off of another (host) organism.

Pendulum—a body suspended from a fixed point so that it is free to swing to and fro.

Pent—closely confined; also used as a prefix indicating five (pentagon).

Pentane (C_5H_{12})—a hydrocarbon liquid (at room temperature) found in crude oil and used as the main component of gasoline.

Perpendicular—exactly upright or vertical; at right angles to the plane on the horizon.

Petal—a colorful leaf-like flower structure that serves to attract pollinators.

pH—the concentration of hydrogen ions of a substance; on a scale with numbers ranging from 1 to 14, low numbers indicate acidity, high numbers indicate alkalinity.

Pharmaceutical Chemicals—chemicals used in the manufacture of drugs.

Pharynx—the part of the alimentary canal between the cavity of the mouth and the esophagus.

Phenomenon—anything appearing or observed that has scientific interest.

Phosphorus—a nonmetallic element used in matches and incendiary devices; a chemical part of protoplasm.

Photosynthesis—the synthesis of carbohydrates from carbon dioxide, water, and inorganic salts, using light as the source of energy and with the aid of chlorophyll.

Physics—the science which deals with matter, energy, motion, and force.

Pie Graph—same as circle graph.

Pistil—the female reproductive organ of a flower; contains the stigma, style, and ovary.

Pitch—(in music) the degree of height or depth of a tone or a sound, depending upon the rapidity of the vibrations by which it is produced.

Pivot Joint—a joint where one bone rolls or rotates over another; e.g., wrist.

Placenta—a vascular organ that lines the uterine wall during pregnancy and regulates the passage of nutrients to the fetus.

Plankton—microscopic plant and animal life found in fresh and salt water.

Plasma—the liquid part of blood and lymph; as distinguished from the suspended elements.

Plate—crustal rock plates contained in the lithosphere.

Platelet—a small, plate-like body in the blood that assists in blood coagulation.

Platypus—an aquatic mammal found in eastern Australia and Tasmania characterized by web feet and a broad bill.

Pole—either end of the axis of a sphere, such as the Earth; either end of a magnet.

Pollen—the fertilizing element of flowering plants, consisting of fine, powdery, and yellowish grains.

Pollinators—things that transport pollen from plant to plant; can be animal or natural (wind).

Positive—measured or moving in a direction of increase, progress, or forward motion; a quantity greater than zero.

Power—work done or energy transferred per unit of time

Precede—to go before in place, time, or importance.

Precipitate—to separate (a substance) in solid form from a solution, by means of a reagent.

Predator—any plant or animal which is dependent upon other plants or animals for its food.

Predict—to foretell; to tell beforehand the actual event.

Prefix—a letter, syllable, or word put at the beginning of another word to change its meaning.

Prehensile—capable of grasping.

Prey—any animal used by another for food.

Process—a continuous action or series of actions which lead to the accomplishment of a result.

Projectile—a fired or thrown object, such as a bullet or a missile.

Protein—a complex, heavy organic compound that contains amino acids as its basic structural unit; contains nitrogen, carbon, oxygen, hydrogen, and, in many cases, sulfur.

Proton—the small particle, with a positive electrical charge, found in the nucleus of an atom.

Protoplasm—the living matter of organisms regarded as the physical basis of life.

Protozoan—A diverse group of eukaryotes (kingdom Protista) that are primarily unicellular.

Proverb—a short saying, expressing a truth in a few words, e.g., "a stitch in time saves nine."

Pupa—the stage in the life of an insect when it is in a cocoon; the stage in an insect's life between larva and adult.

Pulley—one of the simple machines; a small wheel with a grooved rim on which runs a rope for lifting.

Pulmonary Artery—the large artery that sends blood from the right ventricle of the heart to the lungs.

Pulmonary Vein—the large vein that returns oxygenated blood to the left atrium of the heart from the lungs.

Punnett square—a diagram used to make predictions about the genetic outcomes of mating.

Qualities—an essential or distinctive characteristic; something that distinguishes one person from another.

Quantity—a particular or indefinite amount of anything.

Radiation—energy released in the form of waves or particles.

Radioactive—being capable of emitting radiation; resulting from changes in the nuclei of atoms of an element.

Ray—a beam of energy or light.

Reaction—the reciprocal action of chemical agents upon each other; chemical change.

Recycle—to treat or process (used or waste materials) to make suitable for reuse.

Reflected Ray—a ray of light, heat, or sound after reflection.

Refract—to subject a ray of light to a change in direction.

Relate—to have connection or reference.

Renowned—famed; well known for.

Repel—to drive back; to resist.

Replicate—to produce an exact copy or duplicate.

Reproduction—the process of multiplication of living creatures to perpetuate the species.

Research—scientific investigation and study to discover facts.

Residue—the remnant of a process.

Resistance—opposing or acting against; specifically, the opposition presented to the flow of direct current by a material or device

Review—to go over material previously studied.

Richter Scale—the number system used by seismologists used to determine the strength of an earthquake.

Root—1) the part of a plant which seeks nourishment for the whole plant. 2) base word or stem.

Root System—the part of a plant that anchors it underground and absorbs water and nutrients from the soil.

Rotation—the movement of a body on an axis or center.

Scale—the ratio between dimensions shown on a map or drawing.

Schematic—a diagram, plan, or drawing.

Science—the branch of knowledge dealing with a body of facts arranged to show the operation of general laws.

Scientific Method—a method of research in which a problem is identified, relevant data is gathered, and a hypothesis is formulated and tested.

Screening—a reduction in the intensity of radiation on passing through matter.

Segment—a section, portion, or part.

Seismograph—an instrument which records the distance and strength of earthquakes.

Seismologist—a person who studies and analyzes earthquakes.

Seismology—the study of earthquakes, their causes, and effects.

Sepal—a leaf-like structure that protects the petals of a developing flower.

Shoot—a young plant branch which grows out from the main part of a plant; the main part of a plant growing above ground.

Shoot System—the parts of a plant that are above ground; composed of stems and leaves.

Simple Machine—elementary machines used as a basis for more complex mechanisms (includes the lever, pulley, inclined plane, wheel and axle, wedge, and screw).

Simultaneous—two occurrences happening at the same time.

Sirius—the Dog Star, the most brilliant star in the sky.

Skeletal Muscle—the muscle that is attached to bone and moves the skeleton.

Smog—a mixture of smoke (or other pollutants) and fog in the atmosphere.

Smooth Muscle—involuntary muscle tissue in the walls of viscera and blood vessels, consisting of nonstriated and spindle-shaped cells.

Solar Panels—heat-absorbing plates used to capture the sun's energy for use in home and hot water heating.

Solstice—occurrence when the sun is at its greatest distance from the celestial equator.

Soot—very fine black powder which colors smoke; formed when anything is burned.

South Pole—the southernmost point on Earth, located at 90° South latitude.

Species—the major subdivision of a genus; regarded as the basic category of classification.

Sperm—the male reproductive cell.

Spin Axis—the imaginary line through the North and South poles.

Spine—the spinal or vertebral column; backbone.

Spiral—a plane curve generated by a point moving around a fixed point while constantly receding from or approaching it.

Spore—a walled, single-to-many-celled reproductive body occurring in flowerless plants, such as ferns and certain plant-like animals.

Stamen—the male reproductive organ of a flower; consists of a filament that is topped by a pollen-producing anther.

Stem—the structure in vascular plants which carries materials between the roots and leaves; also provides support for the plant.

Sternum—the breast bone.

Stigma—the top of the pistil of a flower.

Stratosphere—upper part of the atmosphere, greater than six miles above the Earth.

Subscript—something written below the level of normal text.

Style—the stem-like part of the pistil of a flower, supporting the stigma.

Subspecies—an additional category within a common species.

Suffix—a letter, syllable, or word put at the end of another word to change its meaning.

Summarize—expressing only the main ideas of a longer document.

Sutton, Walter—coined the term *genes*. Responsible for naming the gene factor.

Symbiosis—the living together of two dissimilar organisms (as in mutualism or parasitism).

Symbiotic—organisms living in symbiosis.

Symbol—something that represents something else by association or resemblance.

System—a group or combination of parts functioning as a whole according to a specific purpose.

Tarantula—large, hairy spider of the family Theraphosidae.

Technology—science of mechanical and industrial arts.

Tendon—a tough fibrous cord that attaches bone to muscle.

Terminal—one end of an electrical current.

Terrestrial—existing on Earth.

Texture—the visual and especially tactile quality of a surface.

Theory—a statement of the fundamental principles of an art or science rather than of the method of practicing it.

Title—the name of a book, poem, or play; a descriptive or distinctive appellation.

Topic—a subject of essay, discourse, or conversation.

Topographic—scientific or physical features of a region.

Transpire—to emit or give off waste matter (watery vapor) through the surface, as of the body or of leaves.

Trend—to stretch in a certain direction or inclination.

Tropical Rainforest—an environment which consists of high temperature, high humidity, and heavy rains.

Turgor—the normal rigidity of plant cells, resulting from the pressure exerted by the cell contents on the cell walls.

Ultraviolet rays—strong invisible light rays, with short wavelengths; ultraviolet rays from the Sun can damage the skin and cause cancer.

Umbilical cord—a fibrous cord, joining the fetus to the placenta of the mother, through which nutrients pass from the mother and waste passes from the fetus.

Universe—the totality of known or supposed objects and phenomena throughout space.

Uranium—a radioactive and metallic element used in photography; the 235 isotope is used for atomic reactions.

Variation—a modification or change; an amount of change or difference.

Vein—a blood vessel that carries blood from the outer reaches of the body back to the heart.

Vernal—belonging to or appearing in the spring.

Vertebra—one of the 33 bones in the spinal column.

Vicissitude—regular change or succession.

Virus—an ultramicroscopic, metabolically inert infectious agent that replicates only within the cells of living hosts (e.g. common cold).

Volt—defined as the difference of electrical potential between two points of a conductor carrying a constant current of one ampere, when the power dissipated between these points is equal to one watt.

Voltage—the total number of volts in a particular electrical circuit; the total number of volts available at the source of electrical power.

Voltmeter—an instrument for measuring the voltage of an electric current.

Voluntary Muscle—those muscles that an individual can control; e.g., leg or arm muscles.

Waterscrew—a simple machine invented by Archimedes.

Watt—an SI unit of power, equivalent to one joule per second and equal to the power in a circuit in which a current of one ampere flows across a potential difference of one volt.

Wave—the oscillating (back and forth) motion by which light and sound are carried.

Wavelength—the distance between two adjacent peaks in the progress of a wave.

Work—force times the distance through which it acts.

Yeast—fungus that reproduces by budding and is capable of fermentation in sugar solutions and starchy substances.

Yucatan—a peninsula in southeast Mexico.

Zygote—a fertilized egg formed by the union of a sperm with the egg.

Science

Requirements for Issuance of Certificate/Diploma

REQUIREMENTS FOR ISSUANCE OF CERTIFICATE/DIPLOMA

Location	Minimum Test Score	Minimum Age For Credential	Residency Requirement	Minimum Age For Testing	Testing Fee Per Battery	Title Of Credential
UNITED STATES						
Alabama	400 min & 450 avg	18[1]	30 days	18[1]	$25.00[2]	Cert. of H.S. Equiv.
Alaska	400 min & 450 avg	18[1]	resident	18[1]	max. $25.00	H.S. Dipl.
Arizona	400 min & 450 avg	18[1]	none	18[1]	max. $25.00[2]	H.S. Cert. of Equiv.
Arkansas	400 min & 450 avg	16	legal resident	16[1]	none	H.S. Dipl.
California	400 min & 450 avg	18[1]	resident	18[1]	varies	H.S. Equiv. Cert.
Colorado	400 min & 450 avg	17	resident[1]	17	$25.00-$40.00	H.S. Equiv. Cert.
Connecticut	400 min & 450 avg	17[1]	resident	17[1]	over 21, $13.00	H.S. Dipl.
Delaware	400 min & 450 avg	18	resident	18[1]	$25.00	St. Bd. of Ed. Endsmt.
District of Columbia	400 min & 450 avg	18	resident[1]	18[1]	$20.00	H.S. Equiv. Cert.
Florida	400 min & 450 avg	18	legal resident	18[1]	$25.00	H.S. Dipl.
Georgia	400 min & 450 avg	18[1]	none	18[1]	$35.00	Gen. Educ. Dev. Diploma

[1] Jurisdictional requirements on exceptions and limitations.
[2] Jurisdictional requirements on credential-related and other fees.

GED Science

Location	Minimum Test Score	Minimum Age For Credential	Residency Requirement	Minimum Age For Testing	Testing Fee Per Battery	Title Of Credential
Hawaii	400 min & 450 avg	17	resident[1]	17[1]	$20.00	Dept. of Ed. H.S. Dipl.
Idaho	400 min & 450 avg	18	resident	18[1]	varies	H.S. Equiv. Cert.
Illinois	400 min & 450 avg	18[1]	30 days	18[1]	$15.00[2]	H.S. Equiv. Cert.
Indiana	400 min & 450 avg	17[1]	30 days	17[1]	maximum $25.00	H.S. Dipl.
Iowa	400 min & 450 avg	17[1]	none	17[1]	$20.00[2]	H.S. Equiv. Dipl.
Kansas	400 min & 450 avg	16[1]	resident[1]	16[1]	$30.00	H. S. Dipl.
Kentucky	400 min & 450 avg	16	resident	16[1]	$25.00	H.S. Equiv. Dipl.
Louisiana	400 min & 450 avg	17[1]	resident[1]	17[1]	maximum $20.00	H.S. Equiv. Dipl.[1]
Maine	400 min & 450 avg	18[1]	none	18[1]	none[2]	H.S. Equiv. Dipl.
Maryland	400 min & 450 avg	16[1]	3 months	16	$18.00[2]	H.S. Dipl.
Massachusetts	400 min & 450 avg	19[1]	resident	19[1]	$40.00	H.S. Equiv. Cert.
Michigan	400 min & 450 avg	18[1]	30 days	16[1]	varies	H.S. Equiv. Cert.
Minnesota	400 min & 450 avg	19[1]	resident	19[1]	$40.00	Sec. Sch. Equiv. Cert.

[1] Jurisdictional requirements on exceptions and limitations.
[2] Jurisdictional requirements on credential-related and other fees.

Requirements for Issuance of Certificate/Diploma

Location	Minimum Test Score	Minimum Age For Credential	Residency Requirement	Minimum Age For Testing	Testing Fee Per Battery	Title Of Credential
Mississippi	400 min & 450 avg	17	30 days[1]	17[1]	$20.00	H.S. Equiv. Dipl.
Missouri	400 min & 450 avg	16[1]	resident[1]	16[1]	$20.00	Cert. of H.S. Equiv.
Montana	400 min & 450 avg	17[1]	resident[1]	17[1]	$18.00	H.S. Equiv. Cert.
Nebraska	400 min & 450 avg	18	30 days[1]	16[1]	$20.00–$30.00[2]	Dept. of Ed. H.S. Dipl.
Nevada	400 min & 450 avg	17	none	17	$25.00[2]	Cert. of H.S. Equiv.
New Hampshire	400 min & 450 avg	18	resident	18[1]	$40.00	Cert. of H.S. Equiv.
New Jersey	400 min & 450 avg	16[1]	resident	16[1]	$20.00	H.S. Dipl.
New Mexico	400 min & 450 avg	18[1]	resident	18[1]	varies[2]	H.S. Dipl.
New York	400 min & 450 avg	19[1]	1 month	19[1]	none	H.S. Equiv. Dipl.
North Carolina	400 min & 450 avg	16	resident[1]	16[1]	$7.50[2]	H.S. Dipl. Equiv.
North Dakota	400 min & 450 avg	18[1]	none	18[1]	varies	H.S. Equiv. Cert.
Ohio	400 min & 450 avg	19[1]	resident	19[1]	$42.00[1,2]	Cert. of H.S. Equiv.

[1] Jurisdictional requirements on exceptions and limitations.
[2] Jurisdictional requirements on credential-related and other fees.

GED Science

Location	Minimum Test Score	Minimum Age For Credential	Residency Requirement	Minimum Age For Testing	Testing Fee Per Battery	Title Of Credential
Oklahoma	40 min & 450 avg	16[1]	resident	16[1]	varies[2]	Cert. of H.S. Equiv.
Oregon	40 min & 450 avg	18[1]	resident[1]	18[1]	varies[2]	Cert. of Gen. Ed. Dev.
Pennsylvania	40 min & 450 avg	18[1]	resident[1]	18[1]	varies	Com. Sec. Sc. Dipl.
Rhode Island	40 min & 450 avg	16[1]	resident	16[1]	$15.00	H.S. Equiv. Dipl.
South Carolina	40 min & 450 avg	17	resident[1]	17[1]	varies	H.S. Equiv. Dipl.
South Dakota	40 min & 450 avg	18[1]	resident[1]	17[1]	maximum $20.00	H.S. Equiv. Cert.
Tennessee	40 min & 450 avg	18[1]	resident	18[1]	$20.00-$25.00	Equiv. H.S. Dipl.
Texas	40 min & 450 avg	18[1]	resident[1]	18[1]	varies[2]	Cert. of H.S. Equiv.
Utah	40 min & 450 avg	17[1]	resident[1]	17[1]	$25.00 and up	Cert. of Gen. Ed. Dev.
Vermont	40 min & 450 avg	16	none	16[1]	$25.00-$30.00	Sec. Sc. Equiv. Cert.
Virginia	40 min & 450 avg	18[1]	resident	18[1]	$25.00[2]	Com. Gen. Ed. Dev. Cert.
Washington	40 min & 450 avg	19[1]	resident	19[1]	$25.00	Cert. of Ed. Comp.
West Virginia	40 min & 450 avg	18[1]	30 days	18[1]	varies	H.S. Equiv. Dipl.

[1] Jurisdictional requirements on exceptions and limitations.
[2] Jurisdictional requirements on credential-related and other fees.

Requirements for Issuance of Certificate/Diploma

Location	Minimum Test Score	Minimum Age For Credential	Residency Requirement	Minimum Age For Testing	Testing Fee Per Battery	Title Of Credential
Wisconsin	400 min & 450 avg	18	voting resident	18[1]	varies	H.S. Equiv. Dipl.
Wyoming	400 min & 450 avg	18	resident[1]	17[1]	varies	H.S. Equiv. Cert.
CANADA - PROVINCES & TERRITORIES						
Alberta	450 min each test	18[1]	resident	18	$50.00	H.S. Equiv. Dipl.
British Columbia	450 min each test	19[1]	resident	19	$26.75	Sec. Sc. Equiv. Cert.
Manitoba	450 min each test	19[1]	resident	19	$22.00	H.S. Equiv. Dipl.
New Brunswick	450 min each test	19	resident	19	$10.00	H.S. Equiv. Cert.
Newfoundland	400 min & 450 avg	19[1]	resident	19	none	H.S. Equiv. Dipl.
Northwest Terr.	450 min each test	18[1]	6 months	18[1]	$5.00	H.S. Equiv. Dipl.
Nova Scotia	450 min each test	19[1]	none	19	$20.00	H.S. Equiv. Cert.
Prince Edward Is.	450 min each test	19[1]	resident	19[1]	$20.00	H.S. Equiv. Cert.
Saskatchewan	450 min each test	19	resident	19[1]	$25.00	H.S. Equiv. Cert.

[1] Jurisdictional requirements on exceptions and limitations.
[2] Jurisdictional requirements on credential-related and other fees.

GED Science

Location	Minimum Test Score	Minimum Age For Credential	Residency Requirement	Minimum Age For Testing	Testing Fee Per Battery	Title Of Credential
Yukon	450 min each test	19[1]	resident	19[1]	$25.00	Sec. Sc. Equiv. Cert.
U.S. TERRITORIES						
American Samoa	400 min & 450 avg	17	resident	17[1]	$20.00	H.S. Dipl. of Equiv.
Guam	400 min & 450 avg	18	resident	18[1]	$10.00	H.S. Equiv. Dipl.
Kwajalein	400 min & 450 avg	18	resident	18	$27.50	H. S. Equiv. Dipl.
Mariana Islands	400 min & 450 avg	18[1]	30 days	18[1]	$5.00[2]	H.S. Equiv. Dipl.
Marshall Islands	400 min & 450 avg	17[1]	30 days	17	$7.50[2]	H.S. Equiv. Dipl.
Micronesia	400 min & 450 avg	18	resident	18[1]	$7.50[2]	H.S. Equiv. Cert.
Palau	400 min & 450 avg	16	Contact your local Dept. of Ed.	16[1]	$10.00	Cert. of Equiv.
Puerto Rico	400 min & 450 avg	18	resident	18	no charge	H.S. Equiv. Dipl.
Virgin Islands	400 min & 450 avg	18	none[1]	17	$20.00	H. S. Dipl.

[1] Jurisdictional requirements on exceptions and limitations.
[2] Jurisdictional requirements on credential-related and other fees.